BASIC CONCEPTS OF NUCLEAR PHYSICS

BASIC CONCEPTS OF NUCLEAR PHYSICS

Jagadish B. Garg

PROFESSOR OF PHYSICS

UNIVERSITY AT ALBANY, NEW YORK

To order additional copies of this book, contact:
Xlibris
1-888-795-4274
www.Xlibris.com
Orders@Xlibris.com
564509

Contents

PREFACE

Nuclear physics is a core subject in physics, and it is quite extensive in its content and importance. This is due to extensive research, which had been carried out by many thousands of scientists all across the world since the discovery of nucleus by Rutherford in 1911 and the vigorous research activity in this field after the discovery of nuclear fission in 1938. This discovery culminated in the making of the atom bomb and later in the pursuit of peaceful uses of nuclear energy all around the world.

The field of nuclear physics was developed by scientists conducting numerous experiments to observe certain facts and later the efforts of scientists to explain these facts by visualization and interpretation with the help of mathematical tools and in particular with the use of quantum mechanics.

To promote the advancement of the field of nuclear physics, U.S. government established various laboratories such as Brookhaven National Laboratory, Argonne National Laboratory, Oak Ridge National Laboratory, and Los Alamos National Research Laboratory. Scientists of many universities conducted research in these laboratories and made important contribution to the field of nuclear physics.

As a consequence, research in the field of nuclear physics was pursued vigorously, and it became the dominant field of research and teaching throughout the world during the years 1950-1985.

At the same time for the pursuit of nuclear energy, many nuclear reactors were installed. A new field of nuclear engineering was established in many universities. Unfortunately, due to few nuclear accidents such as one at Chernobyl in USSR and the other at a three-mile reactor in Pennsylvania in this country, this has raised concerns among the politicians as well as the general public about the safety of nuclear reactors. Consequently, no new reactors were built in the country in the past thirty years. Over the years, the funds for research in the field of nuclear physics became scarce, and it affected the teaching and research of nuclear physics around the world.

Many universities, instead of requiring a full-year course in nuclear physics by graduate students, opted to offer either one-semester survey course in nuclear physics or none for their PhD students pursuing other fields of physics.

Many excellent textbooks in nuclear physics were published during the years 1950-1975, and due to lack of demand for a textbook, many of these books are now out of print.

I have taught a one-semester survey course in nuclear physics for the past twenty years. While teaching this course, I felt the need for a textbook, which could cover the basic topics of nuclear physics in one semester. In recent years, some books have been published combining topics of nuclear physics with particle physics. The research in particle physics has greatly advanced, and many universities are now offering either one semester or a full-year course on this subject. I believe that it is not possible to cover the combined topics of nuclear and particle physics in a one-semester course. Therefore, I have not found such books suited for my survey course in nuclear physics.

The topics covered in my book are nuclear properties, nuclear force obtained from the properties of deuteron and nucleon-nucleon interaction, nuclear structure as discussed in different nuclear models, modes of decay of unstable nuclei, and many types of nuclear reactions and their interpretation. For a one-semester course in nuclear physics, I can only discuss briefly experimental techniques of production and detection of various particles. I have given a brief discussion of application of nuclear physics to some other fields.

I have only quoted references of scientists who are credited with the original ideas or theories, which made important contribution to the field of nuclear physics. I have also quoted references of scientists whose data I have used as illustrations of the concepts of nuclear physics discussed in this book.

This book is written by me alone and is based upon the lecture notes I had prepared from consulting several textbooks on nuclear physics. In my book, I have used significant amount of material from Enge's book Introduction to Nuclear Physics, an excellent book published in 1965 by Addison and Wesley. I would like to acknowledge with thanks the permission given to me by Mrs. Alice William Enge to use this material in my book. I am also thankful to nuclear theorist, Norman Francis, formerly of Knolls Atomic Power Laboratory, for his helpful suggestions, which I have incorporated in my textbook. I will be thankful for any suggestions offered to me by the readers of this textbook. I would like to acknowledge the help of my students, Laxman Mainali, Hassan Mahmood, and Physics Department secretary, Paul LaBate, for their help in typing the manuscript.I would like to thank Muhammad Asim Mubeen for the design of the books front cover.

J. B. Garg

Table of Physical Constants

CONSTANT	SYMBOL	VALUE
Speed of light in vacuum	c	2.997925×10^8 m-$^{sec-1}$
Elementary Charge	e	1.60210×10^{-19} coul
Avogadro constant	N_A	6.02252×10^{23} mole
Electron rest	m_0	9.1091×10^{-31} kg
		$= 5.48597 \times 10^{-4}$ u
Proton rest mass	m_p	$1.67252 \times 10\text{-}27$ kg
		$= 1.00727633$ u
Neutron rest mass	m_n	1.67482×10^{-27} kg
		$= 10086674$ u
Planck constant	h	4.135×10^{-21} Mev. Sec
Plank Constant	h	6.626068×10^{-34}.J. Sec
Compton wavelength of proton	$h/m_p c$	1.32140×10^{-15} m
Bohr magneton	μ_B	9.273×19^{-24} joule-m^2-Wb^{-1}
Nuclear Magneton	μ_N	5.0505×10^{-27} joule-m^{2}Wb^{-1}
Proton moment	μ_p	41049×10^{-26} joule-m^2-Wb^{-1}
Boltzman Constant	k	1.38054×10^{-23} joule-^{0}K^{-1}
Stefan-Boltzmann constatnt	σ	5.6697×10^{-8} W-m^{-2}-^{0}K^{-4}
Gravitational constan	G	6.670×10^{-11} N-m^2-kg^{-2}
One atomic mass	u	931.494 Mev
electron volt	ev	1.60210×10^{-19} joule
Fine Structure constant	e^2/hc	7.2797×10^{-3}
Strong Coupling constant	$\alpha s M_z c^2$	0.119
Length	Fermi	10^{-15} m
Cross section(barns)	b	10^{-24} cm^2

CHAPTER 1

BASIC PROPERTIES OF NUCLEI

1.1. INTRODUCTION

Why Study Nuclear Physics?

Among many disciplines of physics, nuclear physics is considered by many physicists as the most fundamental modern scientific discipline. Even though the field originated at the end of nineteenth century in 1896 with the discovery of radioactivity by Becquerel and the measurement of e/m of the electron in 1899 by J. J.Thomson, the knowledge of nuclear physics advanced very rapidly after the discovery of the nucleus in 1911 by Lord Rutherford. This discovery established the true nature of the atom and allowed Niels Bohr in 1913 to develop the theory of atomic radiation. Next important discovery in this field occurred with the discovery of neutron by James Chadwick in 1932 as the missing particle of the nucleus. The study of neutron induced reaction gave rise to the discovery of nuclear fission of uranium in 1938 by Lise Meitner and O.R. Frisch.This discovery opened doors for the advancement of other fields of physics including solid state physics with the invention of semiconductor devices, computer physics, mathematical physics, and physics of particle accelerators and their detectors.

After the end of the Second World War, the U.S. government established several national laboratories and provided financial support to many universities to build particle accelerators for the study of nuclear physics and for the training of scientists in this field. These activities allowed search for the fundamental nature of nuclear force, nuclear structure, nuclear decay, and nuclear reactions. These studies were pursued by many thousands of physicists all around the world during the period 1940-1980. A great wealth of information was gathered, and thousands of future scientists were trained.

The development of very high energy particle accelerators during 1950-1970 opened the door for the study of meson physics and other elementary particle physics created by high energy nuclear interactions. Scientists then believed that these new particles provided another dimension to the study of nuclear physics. Many nuclear physicists moved to the field of high energy physics, which was later called elementary particle physics. Later, the development of higher and higher energy particle accelerators in trillion electron volts as compared to few million volts in 1930-50 allowed the study of elementary particles' interactions, which advanced our knowledge of nature of nucleon-nucleon force. The field of particle physics gained international importance, and centers of high energy physics were established in United States and Europe. Groups of hundreds of scientists from different countries are collaborating in this field. Many universities are providing a semester-and-a-year course in this field.

There is evidence that the universe was created about fifteen billion years ago with the conversion of energy into matter in a big bang. According to the current knowledge, there are four forces—gravitation, electromagnetic, strong, and weak nuclear force. Scientists now believe that all these forces were unified at the time of the creation of the universe when it was very hot. Elementary particle physicists are now searching ways to unify these forces and have received success in the unification.

Many concepts of symmetries and laws of conservation of energy, linear and angular momentum, parity, charge conjugation, time reversal, and gauge invariance, which were developed in nuclear physics, are vigorously pursued in particle physics.

The field of particle physics has advanced considerably, and its study by students is very essential. However, it would be impossible for an instructor to teach the fundamentals of both fields in one semester course of four-month duration.

Another reason for the study of nuclear physics is that knowledge of nuclear physics is being applied in many fields such as medical, environment, industry, and the production of nuclear energy. In chapter 7 of this book, I have described some of these applications. The reader should read specialized books for more details.

1.2. Rutherford's Discovery

Rutherford and his colleagues, Geiger and Marsden, carried out the experiment using α particles emitted from a radioactive source. In this experiment, a beam of α particle (helium nucleus) with two protons and two neutrons bombarded a gold foil, which contained billions and billions of atoms of gold—each atom carrying seventy nine protons with positive charges inside its nucleus. At the time of this experiment, the general knowledge was that the electric charge was homogenously distributed throughout the entire volume of gold atom. The electric force between similar charges was described by Coulomb's law. This law predicted that α particles, after colliding with gold atoms, will be emitted in the forward directions along the direction of α beam. However, the experiment showed that many α particles were scattered in the backward direction at large angles. This result was surprising and was in conflict with the prediction of a homogenous distribution of charge throughout the atom. The result was in agreement with the prediction that all positive charges in gold atoms were confined in a much smaller volume of sphere with a radius of 10^{-6} times smaller than the radius of the atom.

Rutherford thus concluded that protons in gold atoms were confined in a small radius about 10^{-15} m at the center of the atom. This was called the nucleus of the atom.

Most of the knowledge acquired in the field of nuclear physics is based upon the observation of experimental facts. Study of nuclear physics involves learning about the following:

1. What is a nucleus?
2. What are the building blocks of nuclei? Nucleus consists of particles known as protons and neutrons.
3. How particles interact with each other and with nuclei as a whole.
4. Nature of the nuclear force and nucleon-nucleon interaction.
5. Quantum nature of particles, i.e., mass, charge, angular momentum, spin, magnetic and electric moments, and their energy.
6. Nuclear structure and proposed models of nuclear structure.
7. Decay modes of nuclei-radioactive decay with emission of alpha and beta particles, gamma rays and spontaneous fission.

8. Nuclear reactions between nucleons and nuclei and various models which have been proposed to describe these types of nuclear reactions.
9. Nuclear energy produced in nuclear reactors and nuclear fusion.
10. Development of quark model and the nature of elementary particles.

1.3 What Is a Nucleus?

A nucleus is the central part of an atom. Different nuclei have different number of protons known as atomic number. In 1932, scientist, James Chadwick, (1.2) discovered that a nucleus consists of another particle; he called it a neutron. Neutrons have no charge but have nearly equal mass of protons. Hence, by 1932, nucleus was established as built up of protons and neutrons.

1.4. Nomenclature

A nucleus is represented as X_Z^A, where Z is the atomic number representing number of protons and A is the total mass expressed in atomic mass units of the nucleus. Number of neutrons in the nucleus is $N=A-Z$. Some nuclei may have same Z but different masses. These nuclei are called isotopes such as C_6^{12}, C_6^{13} , C_6^{14} . These are isotopes of carbon. Nuclei with the same value of A and different Z values are called isobars. An example is N_7^{14} O_8^{14} .

1.5. Size and Shape of Nuclei

Nuclei are assumed to have a spherical shape. No one has actually seen the nucleus since it is too small in size. Assuming that the nucleus is spherical, it has a size described by its radius (R) of a sphere. Area of surface of sphere is $4\pi R^2$, and volume of the sphere is $4/3\pi R^3$.

If ρ is the density of the nucleus and m is the mass and v is the volume of the nucleus, then density is given as $\rho = m/v$, which is very large for a nuclei.

Take the nucleus of helium, which has two protons and two neutrons. Mass of each particle is about 10^{-27} kg, and its radius is approximately

1.5×10^{-15}m. Hence, the density of the nucleus is approximately $10^{-27}/10^{-45}$ kg/m^3 or about 10^{18} kg/m^3 as compared to density of water, which is 10^3 kg/m^3. Thus, nucleus is a very dense object.

1.6. Size Measurement

How can one measure the size R of the nucleus? Assuming the nucleus to consist of a number of positively charged protons, one can measure the size of the nucleus by studying the interaction of protons when they come near the nucleus and are scattered by their charge. Similarly, electrons with a negative charge would experience an attraction with positively charged protons of the nucleus, and they would be scattered by the nucleus.

1.7. High-energy Electron Scattering by a Nucleus

Most extensive study of the size of the nucleus was made by Hofstadter and his colleagues (1.3) at Stanford University using high-energy electrons produced in a mile long electron accelerator and measuring the scattering cross section by a number of nuclei with different masses. Coulomb's law of electric forces between two point charges is given as follows:

$$F = k \frac{q_1 q_2}{d^2} \tag{1.1}$$

Where q_1 and q_2 are charges of proton and electron and d is the distance between them, and k is a constant, whose value is 9×10^9 Nm^2/c^2.

For point charges, one can easily calculate the scattering cross section of electrons by the nucleus using Coulomb's law. One obtains an expression for the differential scattering cross section at an angle Θ subtended by the solid angle d $\Omega = 2\pi \sin \Theta$ dΘ is given as follows:

$$\frac{d\sigma}{d\Omega} = k^2 \frac{Z^2 e^4}{v^4} \left(\frac{2\pi \sin\theta \, d\theta}{\sin^4\left(\dfrac{\theta}{2}\right)} \right) \tag{1.2}$$

where Z is the charge of the target nucleus and υ is the velocity of electrons associated with its energy.

For relativistic electron energies, one should use Mott-Schwinger scattering formula.

Hofstadter Experiment

The apparatus used by Hofstadter et al. is shown in figure 1.1.

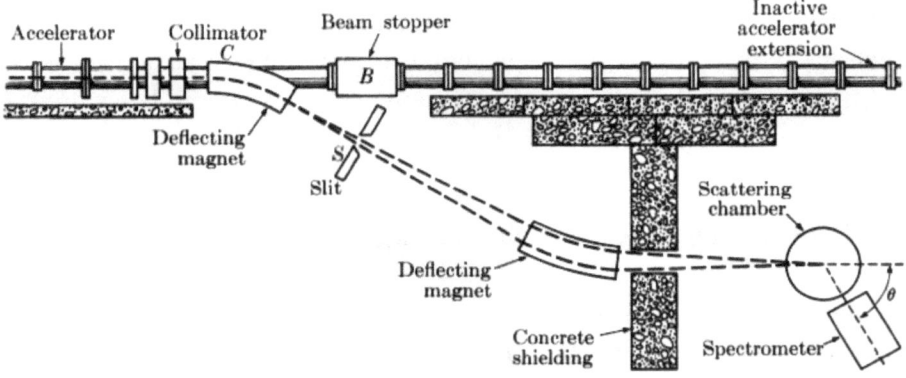

Figure 1.1. Apparatus used by Hofstadter and coworkers for high-energy electron-scattering experiment of nuclei (From R. Hofstadter et al. [1.3])

A beam of high energy (E = 183 MeV) electrons from the accelerator is passed through a collimator and deflected in a scattering chamber containing a target of element gold. The angular distribution of electrons scattered by the target is measured and plotted as shown in figure 1.2.

These plots are for two different electron energies of 153 MeV and 183 MeV. Analytic expression based upon Born approximation is given as follows:

$$\frac{d\sigma}{d\Omega}(\theta) = \left(\frac{Ze^2}{2E_e}\right)^2 \cos^2\frac{\theta}{2}\left[1 + \frac{\pi z}{137}\frac{\sin\frac{\theta}{2}\left(1 - \sin\frac{\theta}{2}\right)}{\cos^2\left(\frac{\theta}{2}\right)}\right] \qquad (1.3)$$

Where θ is the scattering angle of electrons in the center of mass.

In figure 1.2, points are experimental data, and solid curves are best fit for chosen values of R_e and Z.

In their calculation, Hofstadter et al. assumed that the charge distribution inside the nucleus is given as follows:

$$\rho = \rho_0 \frac{1}{1+e^{(r-R_e)/z_1}} \tag{1.4}$$

Where Z_1 is the surface thickness and the radius R_e is the value of R at $r = \frac{1}{2} r$.

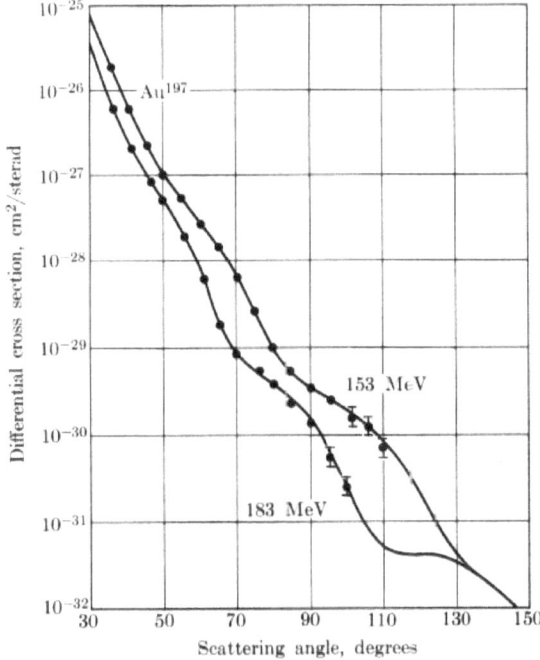

Fig 1.2. Results of electron-scattering experiments performed in gold. (From B. Hahn et al., [1.4]).

The agreement between theory and experiment is good. Hahn et al. (1.4) studied electron scattering in various nuclei with different masses. The measurements on nuclei Ca, V, Co, In, Sb, Au, and Bi are shown in figure 1.3 where the relation between charged density versus radial distance is shown in the figure. These curves show that the charge density is uniform from nuclear center to about 2-3 fermi and then falls off gradually toward the surface. Thickness of the surface is found to be the same for all nuclei. From these measurements, Hahn et al. obtained the values of radii for these nuclei.

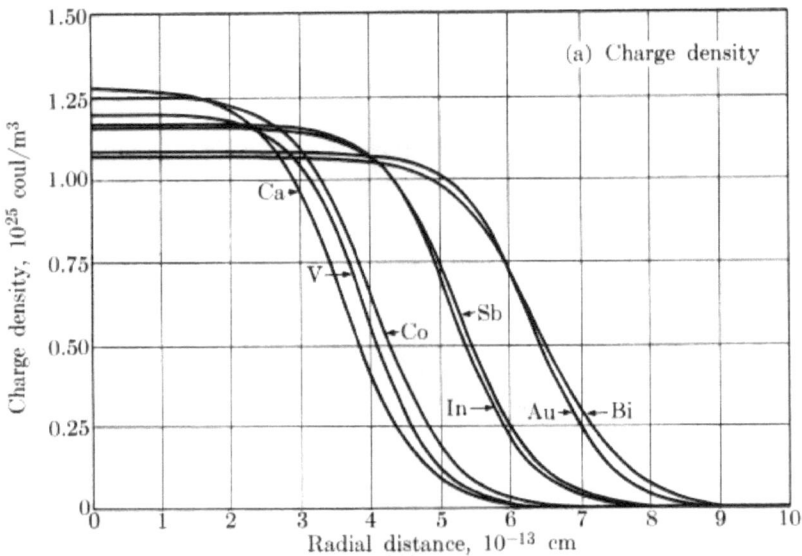

Figure 1.3. Charge density and nucleon density as determined by high-energy electron scattering (From B. Hahn et al., [1.4])

Electron scattering provides values of radii based upon distribution of protons inside the nucleus. A more accurate method of determining distribution of protons as well as neutrons is based upon study of nucleon-nucleon interaction. This can be obtained from the study of interaction of particles such as α particle—protons or neutron from nuclei.

Such measurements usually produce values of radii larger than that obtained from electron-scattering experiments. The nuclear radius **R** is found to be proportional to A where A is the atomic mass and is given as

$$R = r_0 A^{\frac{1}{3}}$$

where r_0 is a constant whose value is obtained from different experiments.

The value of r_0 obtained from different experiments is given below.

Electron scattering   (1.20 ± 0.05) fm

Nucleon scattering   (1.40 ± 0.05) fm

1.8. Nuclear Mass and Binding Energy

Neutrons and protons (n-p) are bound together as a result of nuclear force between these particles. One requires energy to break the neutron and proton apart. This energy, needed to separate the n-p pair, is known as separation energy of the n-p pair. There are many such pairs of n-p in the nucleus, giving rise to total binding energy (BE) of the nucleus.

Hence, BE of the nucleus is the energy that is released when the appropriate number of protons and neutrons are combined to form the nucleus. Binding energy can be calculated for a nucleus by using equation 1.5

$$\text{BE} = \left[M_n - Zm_p - (A - Z)m_n \right]c^2 \tag{1.5}$$

where Z = number of protons, $(A - Z)$ = number of neutrons, m_p = mass of proton, m_n = mass of neutron, and M_n = mass of the nucleus.

1.9. Binding Energy of Nuclei

Scientists calculated the binding energies of many nuclei using the equation 1.5. These total binding energies are divided by the number of nucleons, and one obtains values of binding energy per nucleon. These values in MeV are shown in figure 1.4.

Table 1.1 gives the rest mass energies of some elementary particles.

Particle	(amu)	(Mev)
Electron	5.48597 x 10-4	0.511007
Neutron	1.008665	939.551
Proton	1.007277	938.258
a-particle	4.001506	3727.323

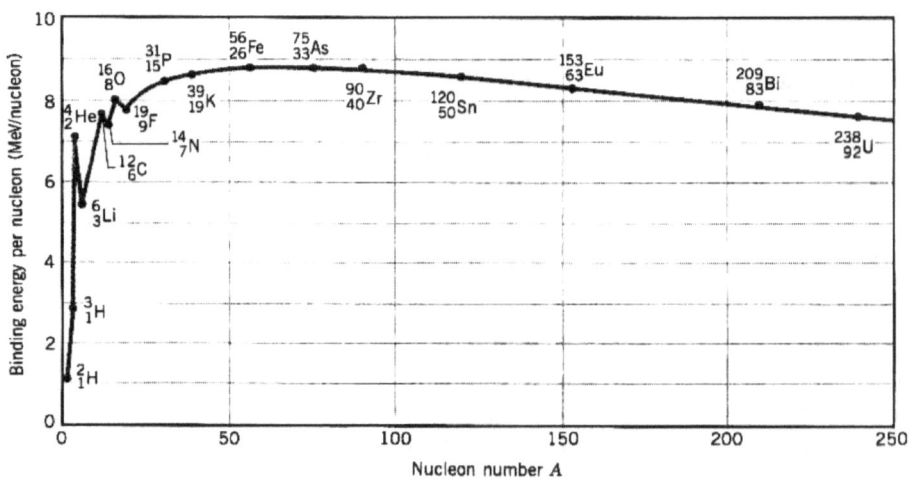

Figure 1.4. BEpernucleon as a Function of Nucleon Number A

From this figure, one sees that BE/nucleon is small 1.1 MeV for deuterium and then gradually increases as a function of nucleon number A, reaching a maximum value of about 8.8 MeV for iron and then gradually decreasing to about 7.5 MeV for A = 250. As will be seen later, the behavior of BE/n plays an important role in determining the stability of nuclei.

1.10. Total Angular Momentum of the Nucleus

One knows that a body or a system of bodies are in a state of constant motion. The nucleus is so small that one cannot see its motion. By intuition, physicists compared the nucleus to the planets in our solar system. From the observation of the planetary motion, one knows that the planets have two types of motion; one is a circular motion in an orbit around the sun caused by the force of gravity between two masses. The other is the spinning motion about its own axis. Borrowing from the planetary motion, scientists have viewed the motion of a nucleus and the nucleons inside it to have similar motions. Niels Bohr used similar ideas for describing the motion of electrons inside an atom.

Angular or Orbital Motion of Nucleons

We know from classical mechanics that when a body rotates with an angular velocity (ω), it posseses an angular momentum. Such a motion of nucleons inside the nucleus gives rise to an angular or orbital momentum designated as l.

In a nucleus, this angular momentum is quantized and has a *value* $\sqrt{l(l+1)}$ in units of $\hbar$. In a frame of spatial coordinate system, this angular momentum vector has components along x, y, and z coordinates. The components along z direction is known as magnetic quantum number m_l, which can take $2l+1$ values. Each nucleon inside the nucleus has an angular momentum l, and for a system of nucleons, these add vectorially to produce a resultant angular momentum (L) of the nucleus.

Spin Angular Momentum of the Nucleons

Nucleons, protons, and neutrons inside the nucleus are also spinning about their own axes. The value of this momentum is also quantized and has a value of ½ units of $\hbar$. These particles are known as fermions, and they obey Fermi-Dirac statistics.

The projection of this momentum along z axis has two components $m_s = +½$ and $m_s = -½$. These represent the direction of spinning motion of the nucleon. For a system of two nucleons, the spin momentum can be combined with the use of Clebsh-Gordan coefficients. If the two nucleons are spinning in the same direction known as parallel motion, the momentum combines to produce a total spin momentum $S = 1$, which has three components along the Z direction and is known to form a triplet state. When the two nucleons are spinning in opposite directions known as antiparallel motion, these produce a total spin momentum $S = 0$ known to form a singlet state. Pauli principle dictates that spin momentum of two identical particles will be $S = 0$ whereas two non-identical particles such as a pair of a proton and a neutron can produce either $S = 0$ or $S = 1$. In a nucleus, there are many such nucleons whose spin moments combine to produce a resultant spin momentum S.

The total spin angular momentum S of all nucleons is added to their total orbital or angular momentum L to produce the total intrinsic spin of the nucleus J given as

$$\mathbf{J} = \mathbf{L} \pm \mathbf{S}$$

where value of J is given in units of ħ as $\sqrt{J(J+1)}$. Total angular momentum J is also called intrinsic spin of the nucleus sometimes denoted by the symbol I.

Component of J along any particular direction such as Z axis (under the magnetic field) is denoted by μ_J known as magnetic quantum number. It can take (2J+1) values as

$$(-J), -J+1, -J+2, \ldots, 0, \ldots, 1, 2, \ldots, (J)$$

1.11. Parity and Symmetry of the Nucleus

Parity is a concept of great importance in atomic and nuclear physics but with no counterpart in classical physics. Behavior of nuclear particles and nuclei are described in terms of a wave function in quantum mechanics. This wave function (ψ) depends upon the coordinates x, y, and z of like particles.

Law of parity is based on the symmetry properties of this wave function. This law says that if the coordinates of a particle are (x, y, z) and if the coordinates are changed to (-x, -y, -z), which is the mirror image of the particle, how the wave function (ψ) will behave.

If ψ(x, y, z) = ψ(-x, -y, -z), i.e., no change ψ has a (+) parity or even parity.

If ψ(x, y, z) = -ψ(-x, -y, -z), i.e., changes sign, it has a (-) or odd parity.

A wave function describing many particles can be written as a product of individual particle wave functions. The parity of the whole nucleus is the product of the single particle wave functions.

From extensive investigations, it has been found that parity is always conserved in strong nuclear interactions. Evidence has been shown that parity is not conserved in nuclear process involving weak interactions such as beta decay.

In dealing with the symmetry properties of wave functions in section 1.13, we discuss the properties of wave functions for an exchange of position of individual particles from -r to +r and exchange of two identical particles.

1.12. Isospin Quantum Number of Nuclei

There is another symmetry property, which applies to exchange of a neutron with a proton or vice versa. Proton has a charge and is therefore subject to Coulomb force whereas neutron has no charge. When one changes a proton with a neutron, the Coulomb force changes the symmetry property of the total wave function of the nucleus. This property can be represented by a new quantum number called the isospin quantum number.

One defines t_3 as isospin projection on a fictitious symmetry axis and assuming that $t_3 = +\frac{1}{2}$ for a proton and $t_3 = -\frac{1}{2}$ for a neutron. Combining this quantum number for all nucleons of a system of nucleons gives the isospin quantum number as $T_3 = \frac{1}{2}(Z-N)$. Thus $T = 0$ for $Z = N$ and $T = 1$ has three values: 1, 0, -1 known as a triplet. Thus, three possible states of two nucleon system will be dineutron, diproton, and deuteron.

1.13. Symmetry Properties of the Wave Function

Another concept in atomic and nuclear physics is that of symmetry. A simple solution of Schrödinger equation for two identical noninteracting particles moving in the same potential is

$$\psi_{n,k} = \psi_n(1)\,\psi_k(2) \tag{1.6}$$

where ψ_n and ψ_k are two solutions of the one-body wave equation and (1) and (2) represent the coordinates of particles 1 and 2. Wave function ψ is not an acceptable solution for two identical particles because it implies that we can label the particles and observe which particles in each state. Wave function can thus be written in the following manner as symmetric and antisymmetric wave functions.

$$\Psi_S = \frac{1}{\sqrt{2}} \ [\Psi_n(1) \ \Psi_k(2) + \Psi_n(2) \ \Psi_k(1) \ ] \tag{1.7}$$

$$\Psi_A = \frac{1}{\sqrt{2}} \ [\Psi_n(1) \ \Psi_k(2) - \Psi_n(2) \ \Psi_k(1) \ ] \tag{1.8}$$

When both particles possess intrinsic spins S, the above equation is multiplied by appropriate spin-wave function.

Experiments have shown that total wave function for a system of particles having $S = 1/2$ is antisymmetric in the exchange of any pair of particles.

1.14. Magnetic Moments of the Nucleus

Nucleus has protons and neutrons. Their motions of charges give rise to currents inside the nucleus. It is known that flow of current gives rise to magnetic field or magnetic moment. Magnetic dipole moment of the nucleus is given as

$$\mu = IA$$

where I is the current and A is area of the loop produced by the orbital motion. A revolving charged particle gives rise to a magnetic moment given as

$$\mu = \left(\frac{ev}{2\pi r}\right)(\pi r^2) = \frac{evr}{2} \tag{1.9}$$

where v is the particle velocity and r is the nuclear radius.
The angular momentum (l) is given as follows:

$$l = mvr$$

The component of magnetic moment along Z direction is given as

$$\mu_z = \left(\frac{e}{2m}\right)\hbar m_l \tag{1.10}$$

Magnetic dipole moment of proton and neutron has been calculated. These are

μ_p = 2.7927 nuclear magnetons
μ_n = -1.9135 nuclear magnetons
1 nuclear magneton (nm) = 5.0505 × 10^{-27} J-m²/Wb

1.15. Electric Moments of Nuclei

Similar to magnetic moments, a nucleus consisting of charged particles can give rise to electric moments. Protons and neutrons are assembled inside the nucleus and give rise to a charge distribution, giving rise to charge density.

Electric moments produced by the nuclear charge currents give rise to dipole and quadrupole moments. Electric dipole moment p_x for an assembly of charges is given as

$$p_x = \int q_i x dx \text{ where } q_i \text{ is the charge}$$

For a point charge, the electric potential at a given distance from the charge is

$$V = k\frac{q}{r}$$

For a charge distribution throughout the volume of the nucleus with radius R, the potential at a distance r from the center is given as

$$V(r) = k\int_v \frac{\rho(r)}{R-r} d\tau \tag{1.11}$$

where ρ is the charge density and $d\tau$ is the element of volume integration, which is from 0 to R. For small value of r/R, the above expression can be expanded in Taylor's series, which is given as

$$V(r) = k\int \rho(r)d\tau + k\int \rho(r)r\cos\theta + \frac{k}{2}\int \rho(r)(3\cos^2\theta - 1)r^2 d\tau \quad (1.12)$$

where z is $r\cos\theta$, the first term is monopole moment, the second term is dipole moment, and the last term is quadrupole moment.

For a symmetrically spherical distribution of charges, all multipole moments vanish. It is found that some nuclei are not spherical but are significantly deformed. The quadrupole moments for such nuclei can take large values.

This topic will be further discussed in later chapters.

Chart of Nuclei

A convenient way to present data concerning the nuclides is offered by a chart of nuclei (prepared by Knolls Atomic Laboratory), which shows the symbol of a nucleus with isotopes and their % abundance, nuclear charge, its atomic mass, and whether the nucleus is stable or unstable, and if unstable, its modes of decay, lifetime of decay, and cross section for thermal neutron capture. One can thus get valuable information about a nucleus by looking at this chart.

1.16-Stability of nuclei

A nucleus consists of a number of protons and neutrons. If protons did not have similar electric charges which produces repulsive force between them, the most stable nuclei will be with equal number of protons and neutrons that is Z= N.

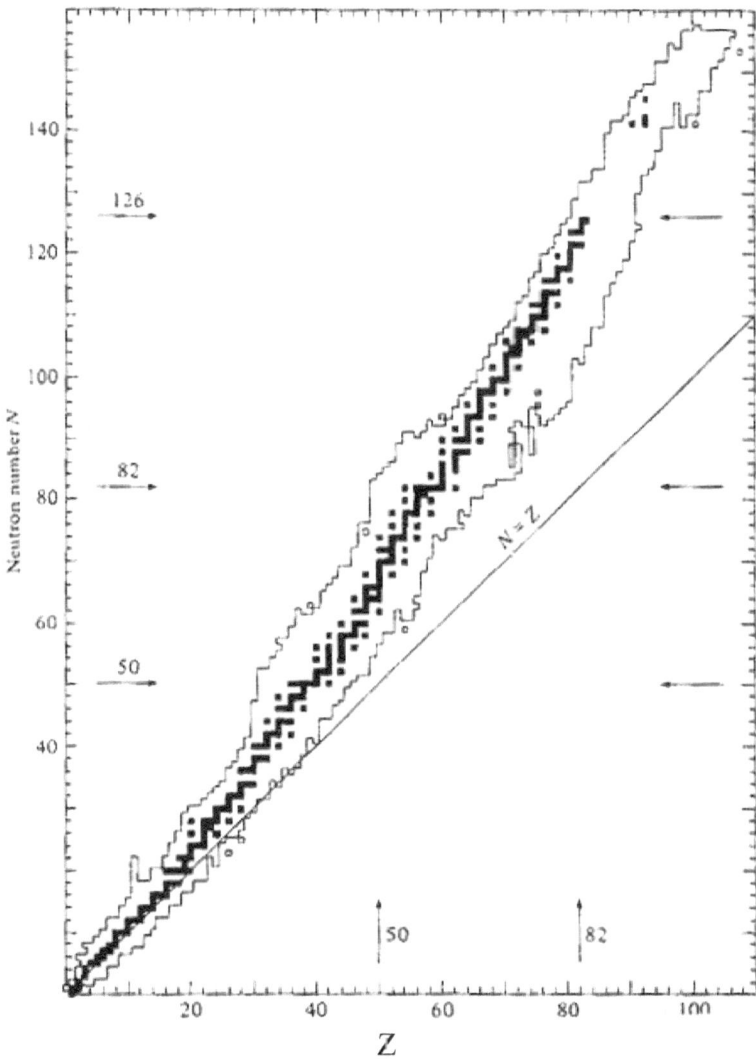

Figure (1.5) shows the distribution of nuclei with number of protons (Z) versus number of neutrons (N).

In this figure one sees that most of the nuclei fall above the N= Z straight line. The nuclei with full squares are the stable nuclei. Nuclei with neutron and proton numbers of 2, 8, 20, 50, 82 and 126 are very stable.

The most stable nuclei are $_2He^4$ and $_{82}Pb^{208}$. Beyond $_{83}Bi^{209}$ nuclei tend to become dynamically unstable and decay by emitting either a α-particles or undergoing spontaneous fission

In general a nucleus becomes unstable if the binding energy of a proton inside a nucleus is more than the binding energy of a neutron, then the proton will convert into a neutron emitting β^+ particle. If the binding energy of a neutron is greater than the binding energy of a proton, the neutron will change into a proton emitting a β^- particle. Such nuclei are unstable against β decay. Detailed discussion of decay of unstable nuclei is given in chapter 5.

1.17. Nuclear Disintegration Processes

A nucleus can be unstable and can decay by various processes. Some of these are shown below.

1.   Alpha (α -) decay
2.   Beta ($\beta^\pm$) decay
3.   Spontaneous fission
4.   Gamma (γ) decay
5.   Electron capture
6.   Internal conversion

Discussion of the above processes is given in chapter 5.

1.   Alpha Decay

All nuclei heavier than lead Pb_{82}^{208} are unstable. Some of these nuclei decay by emitting an α particle (He_2^4) nucleus. Alpha (α) particle (He_2^4) is a tightly bound nucleus and may exist as an entity inside the nucleus.

2.   Beta Decay

This involves decay of the unstable nucleus by the emission of either an electron (e^-) or a positron (e^+). Decay is also accompanied by emission of a neutral particle—a neutrino or an antineutrino. This decay usually takes place in nuclei with excess number of neutrons or protons that are allowed by energy considerations of stability of the nucleus.

3. Spontaneous Fission

Some heavy nuclei may undergo spontaneous fission. In this process, a nucleus breaks up into two or possibly three large fragments and some neutrons. These fragments may also disintegrate by other modes of decay.

4. Gamma Emission

We know that as atoms can be excited to certain higher states and then emit light, similarly nucleus can be excited to higher states, which then emit radiations known as gamma radiation (γ - rays). A gamma ray is a photon with energy $E = h\upsilon$ and has momentum $p = \dfrac{h\upsilon}{c}$. Another decay mode known is internal conversion, which competes with gamma emission.

5. Electron Capture

In this process, an atomic electron from K,L, or M shell is captured by a charged proton, which then emits a neutrino.

6. Internal Conversion

When a nucleus is excited whose decay by gamma emission is prohibited due to selection rule, the excited state energy is given to an atomic electron, which is emitted from the atom.

1.18. Nuclear Reactions

In addition to the study of natural decay of the nucleus, much information can be gained about the nuclear forces and nuclear structure by studying the nuclear reactions.

Particles such as n, p, d, t, α and heavy nuclei can produce different reaction products after colliding with nuclei. A beam of particles with well-defined energy is produced in a particle accelerator and then made to strike a nucleus. Neutrons have no charge and hence cannot be accelerated in a particle accelerator. These can be produced in a nuclear reactor or from a radioactive source.

Some of the Nuclear Reactions Are Described Below

Elastic and Inelastic Scattering of Particles

Interaction of a given particle with a nucleus can also produce its scattering, either elastic or inelastic. In elastic scattering, the energy of the scattered particle is same as that of incident particle. Inelastic scattering causes the energy to change, i.e., scattered particle has less energy. The loss of particle energy is imparted to the target nucleus left in an excited state, which then emits gamma rays.

Transmutation of Nuclei

In addition to scattering of particle, a nuclear reaction can produce other particles and create new nuclei. A typical example of a nuclear reaction is when a proton accelerated in a particle accelerator strikes a target nucleus such as magnesium Mg_{12}^{26} ; it emits an alpha particle and produces sodium as residual nucleus. This reaction is shown below.

$$p_1^1 + Mg_{12}^{26} \rightarrow \alpha_2^4 + Na_{11}^{23}$$

This reaction is usually written as follows:

$$Mg^{26}(p,\alpha)\,Na^{23}$$

In all nuclear reaction, certain quantities are conserved such as charge, mass energy, angular momentum, parity, etc., and is used to predict the nature of nuclear reaction.

Radiative Capture

A typical example is given below. In this reaction, a proton is captured by the nucleus and a γ-ray is emitted.

$$Mg^{26} + p \rightarrow \gamma + Al^{27}$$

Photodisintegration

The reverse reaction produced by the absorption of γ-ray by Al ^{27} can also take place such as

$$\gamma + Al^{27} \rightarrow p + Mg^{26}$$

In this reaction, a gamma ray interacts with the aluminum nucleus and ejects a proton and forms a new nucleus of magnesium.

Induced Fission

A heavy nucleus such as U^{235} after absorption of a neutron can undergo fission. This reaction is known as induced fission and results in the breakdown of the nucleus into several fragments of different nuclei and additional neutrons.

Detailed discussion of nuclear reactions is given in chapter 6.

1.19. Nuclear Energy Levels

One would study that all nuclei have quantum states to which these nuclei could be excited by absorption of energy from nuclear reactions. Nuclei have well-defined energy levels, and each level is characterized by its energy expressed in MeV, total angular momentum J, a parity (±), and its lifetime of decay. Since all the excited states are unstable and must decay to the lowest energy state. A typical example of the excited states of nucleus of N^{14} compiled by Lauritsen et al. (1.5) is shown in figure 1.5.

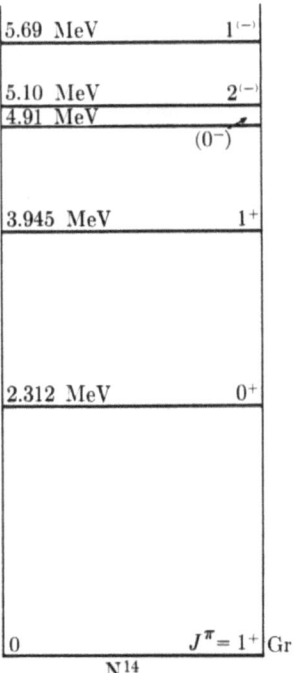

Figure 1.5. The first few excited states of the nuclide $_7N^{14}$ (From T. Lauritsen and Fay Ajzenberg-Selove, "Energy Levels of Light Nuclei," Nuclear Data Sheets, 1962)

Contrary to the atomic case where the excited states of atom are related to the electromagnetic interaction, nuclear excited states are produced by the nuclear interaction and are predicted by the structure of the nucleus based upon different models proposed by scientists. These models of nuclear structure will be discussed in chapter 4.

1.20. Mirror Nuclei

When two nuclei have the same number of nucleons, and the number of protons in one nuclei is the same as number of neutrons in the other nuclei, such nuclei are known as mirror nuclei. Examples are pairs of (Li_3^7, Be_4^7), (Be_4^9, B_5^9), (C_6^{13}, N_7^{13}) .

Energy level diagrams for mirror nuclei have been found to be very similar as are shown in figure 1.6 for C^{13}, N^{13} mirror nuclei.

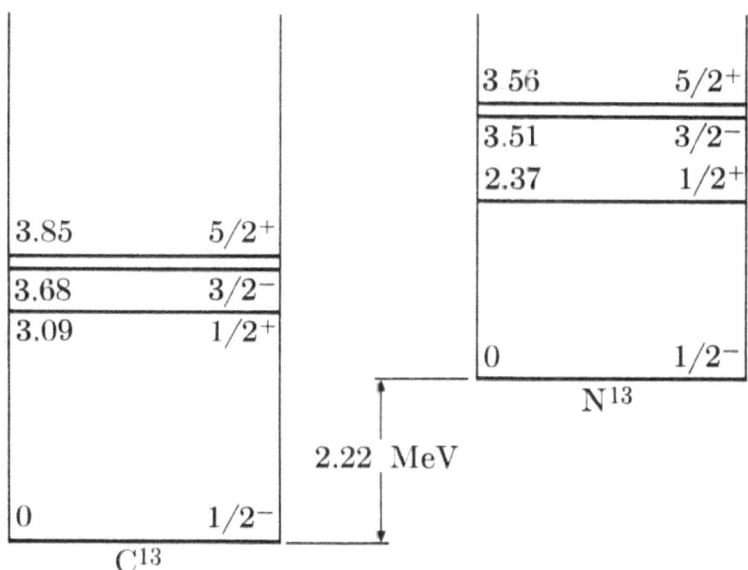

Figure 1.6. Energy level diagram for the mirror pair C^{13}-N^{13}

N^{13} is heavier than C^{13} by an energy equivalent of 2.22 MeV. N^{13} is unstable and decays into C^{13} by emission of a positron. Both nuclei have similar energy levels. Difference in energies is due to difference in electrostatic energy of different number of protons.

For Z protons uniformly distributed inside a spherical volume of radius R, one finds total electrostatic energy as

$$E_c = \frac{3}{5} \frac{Z(Z-1)e^2}{4\pi \epsilon_0 R} \qquad (1.12)$$

whereas total electrostatic energy for a nucleus with Z-1 protons is

$$E_c = \frac{3}{5} \frac{(Z-1)(Z-2)e^2}{4\pi \epsilon_0 R} \qquad (1.13)$$

Taking the difference of electrostatic energy and the difference of masses of neutron, proton, and electron gives the difference in the energy of mirror nuclei as follows:

$$\Delta E_c = \frac{6}{5} \frac{(Z - \frac{1}{2})e^2}{4\pi \in_0 R} - (m_n - m_H)c^2 \qquad (1.14)$$

R is the nuclear radius given as follows:

$$R = r_0 A^{1/3}$$

$r_0 = 1.28$ fm is obtained from electron scattering measurements. Final value of the energy difference between mirror nuclei is

$$\Delta E_c = 1.15(Z - \frac{1}{2})A^{-\frac{1}{3}} - 0.78 \ \text{MeV} \qquad (1.15)$$

$$\Delta E_c = 2.40 \ \text{MeV}$$

This is in good agreement with the experimental value. Average difference between individual levels fluctuates whereas J^π of levels in both nuclei are the same. The fluctuations are small—few tenths of MeV. Accuracy is limited due to certain assumptions made in the above calculation.

CHAPTER 2

NUCLEAR FORCE FROM DEUTERON PROBLEM

Nucleus contains protons (positively charged) and neutrons (neutral charge). These particles are held together inside the nucleus as a result of an attractive force between these nucleons.

Before the discovery of the nucleus, we knew about two forces: one gravitational force (Newton's law of gravitation) and electric force (Coulomb's law). The electric force between two protons having similar charge would be repulsive and would not be able to bind two protons. Similarly, the electric force between a positively charged proton and a neutral neutron would be zero.

We get some idea about the strength of nucleon-nucleon force by the amount of force needed to separate a pair of nucleons from inside the nucleus. For example, the energy equivalent needed to break a pair of neutron and proton from deuteron is about 2.2 MeV and to separate a proton or neutron from a heavy nucleus is about 8 MeV.

The strength of a force is related to its potential energy. For example, the potential energy of a mass (m) at a height (h) above the surface of the earth due to gravitation force is mgh, where g is acceleration due to gravity. Force on mass (m) due to mass of earth is

$$F = \frac{GM_E m}{r^2} = mg$$

where the gravitational constant $G = 8.77 \times 10^{-11} \dfrac{Nm^2}{kg^2}$

Hence, a measurement of potential energy and height of a given mass will provide the value of the magnitude or strength of the force. The magnitude of gravitational force is very small and cannot account for the strength of nuclear force. Similarly, electric Coulomb force cannot account for the nuclear force.

It is therefore obvious that the nuclear force is a new type of force quite different from the well-known electric and gravitational force.

One also knows that nuclear force will be short range, effective only at a very short distance, since the size of nucleus is about 10^{-14} meters. This is quite different from the gravitation and electric forces, which are long-range forces, which get weaker at a large distance, but their strength extends to infinity.

2.1. Force between Two Nucleons

Knowing the value of binding energy of deuteron and the short range nature of the force, Japanese scientist, Hideki Yukawa (2.1), in 1935 proposed the nature of this nuclear force. According to Yukawa, the force was generated by a force field of an unknown particle, which he named π-meson. This particle oscillates between the pair of nucleons. This mechanism is shown in the figure below

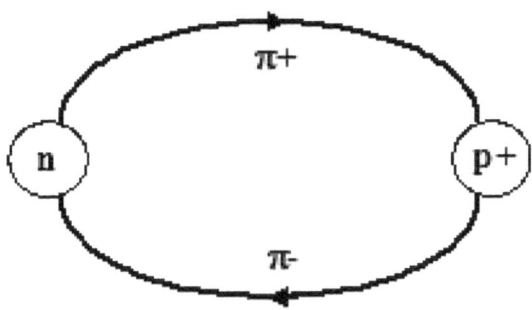

and can be expressed as

$$p^+ \rightarrow n^0 + \pi^+$$
$$n^0 \rightarrow p^+ + \pi^-$$

$$(2.1)$$

Thus, each nucleon acts as a source and a sink for pions, i.e., it is capable of emitting or absorbing them. The π-meson was discovered in 1947 from cosmic rays by Powell and Occhialini (2.2). In 1951, π-mesons were also produced at laboratories with the help of a high-energy particle accelerator.

When two nucleons are brought together, a π-meson emitted by one particle is absorbed by the other with the result that the energy of the system is lowered, and the attractive force between nucleons is generated.

According to Yukawa (2.1), in a neutron-proton interaction, a virtual particle π-meson is emitted. The intrinsic spin of a π-meson is 1, and intrinsic spins of neutrons and protons is ½. Therefore, spin is conserved in this exchange of particles.

Charge is also conserved in this exchange. However, rest energy is not conserved since rest masses of proton, neutron, and π-meson are given below as follows:

$$M_p = 938.26 \text{ MeV}$$
$$M_p = 939.55 \text{ MeV}$$
$$\Delta m = (m_p - m_n) = \pm 1.29 \text{ MeV}$$

The mass of a neutron is larger than the mass of the proton by a small amount of 1.29 MeV whereas the suggested mass of a π-meson is about 140 MeV. Hence, the emission of a π-meson from either a neutron or a proton violates the law of conservation of rest mass energy.

In order to resolve this energy violation, one invokes the law of Heisenberg's principle of uncertainty, which states that the uncertainty in energy is related to the uncertainty in time during which energy is violated. It is given as follows:

$$\Delta t . \Delta E > \hbar \qquad (2.2)$$

The $\hbar$ is Planck's constant, $\Delta E = Mc^2 = 140$ MeV, rest mass of meson, and Δt is time interval of energy violation.

This means that according to uncertainty principle, the energy conservation can be violated for a certain amount of time given as

$$\Delta t = \hbar / \Delta E \qquad (2.3)$$

Substituting the value of $\hbar$ and ΔE, one gets

$$\Delta\Delta t = \frac{1.05\times10^{-34}\,J.S}{(1.6\times10^{-19})(140\times10^{+6})}$$

$$\Delta t = 0.46 \times 10^{-23}\,Sec \tag{2.4}$$

One can determine the range (r) of nuclear force using

$$r = c\Delta t \tag{2.5}$$

where c is the velocity of light assuming that π-meson is traveling with velocity of light; so using equation (2.5), range is given as

$$r = (3\times10^{8})(0.46\times10^{-23}) = 1.38\times10^{-15}\,m \tag{2.6}$$

where 1.38 fm is nuclear dimension or range of nuclear force if it is caused by π-meson of mass $\sim 140\ MeV/c^2$.

2.2. Modern Theory of Nucleons

As a result of the discovery of many new particles, Murray Gell-Mann (2.3) and George Zweig independently proposed that all particles hadrons and mesons were not elementary particles, but these were made up of smaller and more elementary particles. Gell-Mann called these quarks, and Zweig called these aces.

In the beginning, Gell-Mann proposed two quarks and named them up and down quarks; however, as time passed, four more quarks were discovered as resonances in the interaction of very high energy protons with nuclei.

One of the main characteristics of these quarks is that these carry fractional units of charges.

The table below shows the names, charges, and masses of these quarks. All quarks have spin ½ and have antiquarks with charges opposite to the charges of quarks.

Based upon the charges of proton and a neutron, Gell-Mann proposed that a proton is made up of 3 quarks (uud), and a neutron is made up of 3 quarks (udd).

Table 1.1. Quarks, their names, symbols, charges, and masses

Name	Symbol	Charge	mass (Mev/c ^{2})
Up	u	+ 2/3	330
Down	d	- 1/3	333
Strange	s	- 1/3	486
Charm	c	+ 2/3	1650
Top	t	+ 2/3	4500
Bottom	b	- 1/3	770,000

Thus, according to Gell-Mann theory, the force between a pair of neutron-proton is replaced by the force between quarks. Modern theory assumes that the force between quarks is caused by a massless particle called gluon. Though gluon has not been observed in isolation, in very high energy (p-p) collisions, jets of gluons being emitted from the collision are observed.

The topic of quark theory and their interaction with nuclei are discussed in books on elementary particles. In this book, I will follow the discussion of more established theory of nuclear f orce based upon the known properties of deuteron.

2.3. Nuclear Force Determined from the Nucleus Deuteron

Nucleus of deuteron contains one proton and one neutron. This is the lightest, stable two nucleon system. Hence, its properties will shed light on the nuclear force between a pair of a neutron and a proton n-p.

Experiments have provided the following properties of deuteron:

binding energy = -2.2244 MeV
total angular momentum and parity $J^{\pi} = 1^{+}$
root mean square radius (size) of the deuteron $\sqrt{r_d^2} = 2.1$ Fermi
magnetic moment $\mu_d = 0.857393$ nm
electric quadrupole moment $Q = 0.00282$ barn

A theory based upon quantum mechanics is applied here. Schrödinger equation describing two body problem in the center of mass system is written as

$$-\frac{\hbar^2}{2m}\Delta^2\Psi + V\Psi = E\Psi \tag{2.7}$$

Here, m is the reduced mass of the n-p system = $\dfrac{m_1 m_2}{m_1 + m_2}$

where m_1 and m_2 are masses of proton and neutron respectively. V is the potential energy related to the nucleon-nucleon force, and E is the binding energy of the deuteron, which is equal to negative 2.225 MeV.

It is assumed that the force (F) between the pair of nucleons is short range and is attractive.

$$F = -\frac{d(PE)}{dr} \tag{2.8}$$

where PE is the potential energy.

For a spherically symmetrical potential, the wave function Ψ can be separated as a radial function u(r) and $Y_{lm}(\theta,\phi)$ known as spherical harmonic function. That is given as

$$\Psi = \frac{u(r)}{r} Y_{lm}(\theta,\phi)$$

For $\ell = 0$, $Y_{lm}(\theta,\phi) = (4\pi)^{-\frac{1}{2}}$ and the wave function is given as

$$\Psi = \frac{u(r)}{r}$$

Substituting $\dfrac{u(r)}{r}$ in the equation 2.7, one has for $\ell = 0$

$$-\frac{\hbar^2}{2m}\frac{d^2u(r)}{dr^2}+(V-E_B)u(r)=0$$

$$\frac{d^2u(r)}{dr^2}+\frac{2m}{\hbar^2}(E_B-V)u(r)=0$$

since $V, and \ E_B$ are negative.

For $\ell > 0$, one has an additional term $\dfrac{l(l+1)\hbar^2}{2mr^2}$, which is the centrifugal potential and the wave function for $\ell > 0$ becomes

$$\frac{d^2u(r)}{dr^2}+\frac{2m}{\hbar^2}(E-V-\frac{l(l+1)\hbar^2}{2mr^2})u_l(r)=0 \qquad (2.9)$$

The centrifugal potential is plotted in figure 2.1 as a function of distance r (fm) for the neutron-proton in the center of mass.

It is zero for $\ell = 0$. Such potential is repulsive, and it causes the particles to move apart from each other.

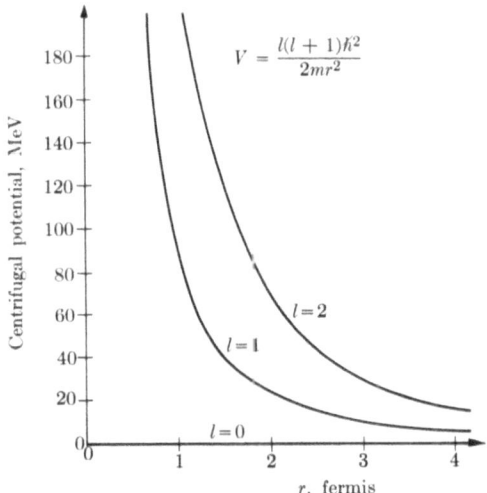

Figure 2.1. Centrifugal potential for neutron-proton system

2.4. Simple Theory of the Deuteron

We will treat the problem for $\ell = 0$ case. The radial function is determined by using a nuclear potential as a function of distance r. The potential used in this analysis is a square-well potential given as follows:

$$V = \infty \quad r < c \qquad\qquad ....(1)$$

$$V = -V_0, c < r < b \qquad ....(11)$$

$$V = 0, \quad r > b + c \qquad ....(111)$$

A square-well potential may not be the actual potential, but it is used here for simplicity. Other forms of potentials such as Gaussian, exponential, and Yukawa have been used by scientists, but all forms of potential give similar results except the strength and range values are slightly different.

Since the two nucleons cannot come closer than a certain distance, it is assumed that they experience a repulsive force for distance less than c known as hard core infinite repulsive potential. This extends from center of the nucleus to the distance c as shown in figure 2.2.

Two nucleons thus make a bound system only in region where the force is present. The wave function behaves exponentially for $r > b + c$. In region II where the potential is large and negative, the possible solution of the wave function is shown in the following:

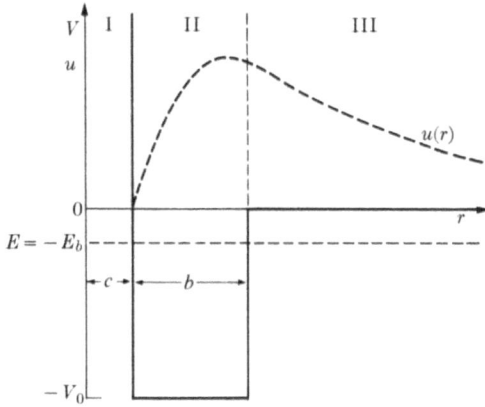

Figure 2.2. Simplified neutron-proton potential and deuteron radial wave function u(r)

$$u(r) = A\sin K(r-c) \ \ for \ r < b + c \tag{2.10}$$

where the wave number K is given as

$$K = \frac{\sqrt{2m(V_0 - E)}}{\hbar} \tag{2.11}$$

In region III outside the range of nuclear potential, the accepted solution of the wave equation is

$$u(r) = Be^{-kr} for \ r > b + c \tag{2.12}$$

where A and B are constants, whose values are determined later, and k is given as

$$k = \frac{\sqrt{2mE}}{\hbar} \tag{2.13}$$

Since the radial wave function has to be continuous at the boundary at r = b+c, it follows that the wave functions at the boundary match or

$$A\sin K(b) = Be^{-k(c+b)} \tag{2.14}$$

and their first derivative must also match as

$$KA\cos K(b) = -kBe^{-k(c+b)} \tag{2.15}$$

dividing (2.15) by (2.14) one gets

$$\frac{KA\cos K(b)}{A\sin K(b)} = \frac{-kBe^{-k(c+b)}}{Be^{-k(c+b)}} \tag{2.16}$$

$$or \ K\cot Kb = -k \tag{2.17}$$

This equation gives a relation between the nuclear potential V_o, binding energy of deuteron E_B, and the range b of the nuclear potential

$$\cot Kb = -\frac{k}{K} \tag{2.18}$$

$$b = \frac{1}{K} Arc \cot(-\frac{k}{K}) \tag{2.19}$$

Assuming a value for $V_o = 40$ MeV much greater than the deuteron's binding energy and $m = 0.504$ u, then one obtains values of K and k as

$$K = \frac{\sqrt{2mV_0}}{\hbar} = 0.955 F^{-1}$$

and $k = 0.0155\sqrt{2.22}F^{-1}$

or k = 0.232 F^{-1}

and $b = \frac{1}{0.955\,F^{-1}} Arc \cot\left(\frac{-0.232}{0.955}\right)$

or $b = 1.895$ F as the range of nuclear potential.

By choosing different values of V_o from 20 MeV to 100 MeV, one gets different values for b. Therefore, the range of the force depends upon the nuclear potential. This treatment does not provide unique values of both strength and range of the nuclear force. It gives only a value for the product of V_o and b.

However, further information for V_o and b can be obtained from the known value of deuteron's radius. We know from 2.18 that

$$\cot Kb = -\frac{k}{K}$$

Substituting value of K and $k,$ one gets

$$\cot\frac{\sqrt{2m_0\left(V_0-E_B\right)}}{\hbar}b=-\frac{\sqrt{2mE_B}}{\sqrt{2m(V_0-E_B)}} \tag{2.20}$$

$$\cot\frac{\sqrt{2m\left(V_0-E_B\right)}}{\hbar}b=\frac{\sqrt{E_B}}{\sqrt{(V_0-E_B)}}$$

For an unbound system $E_B = 0,$ substituting values of $m = 0.0504\ u$ and $\hbar$ in the equation 2.20, one gets

$$\cot\frac{\sqrt{2mV_0}}{\hbar}b=0$$

$$\frac{\sqrt{2mV_0}}{\hbar}b=\frac{\pi}{2}$$

$$V_0b^2=\frac{\pi^2\hbar^2}{8m}$$

$$V_0b^2=102MeV\text{ - }F^2 \tag{2.21}$$

where F is in fm. This functional relationship between V_0 and b given by 2.18 is plotted in figure 2.3 as curve a.

Curve b is for unbound system $(E_B = 0)$ given by 2.21.

If one accepts the value of root mean square radius of 2.1 fm for deuteron as determined from electron scattering and the value of repulsive core $c = 0.4\ fm$ as determined from high energy electron scattering, one gets another relationship of V_0 versus b as shown by the curve c in figure 2.3.

The intersection of this curve with curve a gives a value of the well depth of 73 MeV and the range b = 1.33 fm.

These values are consistent both with the binding energy and the root mean square value of deuteron radius.

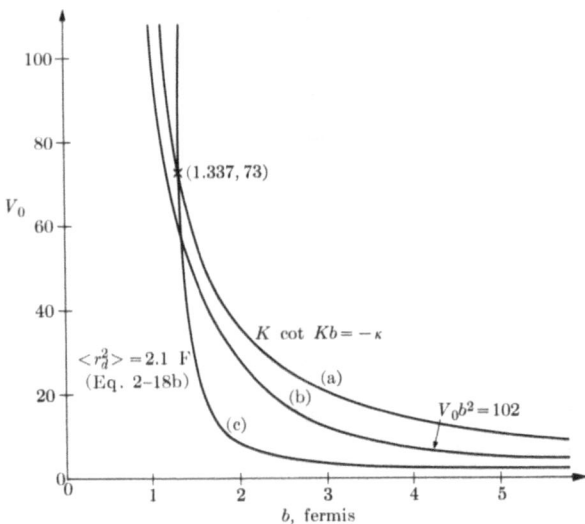

Figure 2.3. Potential-well depth V_0 versus well width b

a gives correct binding energy for deuteron b, which barely binds a neutron to a proton, and c which gives correct size of deuteron.

2.5. Deuteron's Radius

The measurement of deuteron's radius was carried out at Stanford University's electron linear accelerator by McIntyre and colleagues (2.4).

They determined the root mean square radius of deuteron as 2.1 F. The *rms* radius of proton was approximately 0.8 F since the electron scattering method determines the charge distribution of the nucleus. This value may be somewhat smaller than the actual radius of deuteron, which consists of a proton and a neutral particle neutron.

One can determine theoretically the root mean square radius of deuteron as follows:

2.6. Normalization of the Deuteron Wave Function

The constants A and B of the radial wave function in equation 2.14 were left undetermined since values of V_o and b do not depend upon them.

However, one needs to know their values in order to determine the *rms* radius of the deuteron.

Total probability of wave function Ψ is 1.

$$\psi = \frac{1}{r} u(r) Y_{l.m}(\theta, \phi)$$

$$For\ l = 0,\ Y_{l.m}(\theta, \phi) = \frac{1}{\sqrt{4\pi}}$$

Total probability for the deuteron to be in an element of volume $(d\tau)$ is given as (spherical shell of thickness dr) is

$$\int |\Psi|^2 d\tau$$

where $d\tau = 4\pi\ r^2 dr$, r being the radius of a sphere.

$$\Psi^2 = \int_0^\infty \left| \frac{u(r)}{r} \frac{1}{\sqrt{4\pi}} \right|^2 d\tau$$

$$\Psi^2 = \int_0^\infty \left| \frac{u(r)}{r} \right|^2 \frac{1}{4\pi} 4\pi r^2 dr = 1$$

$$\Psi^2 = \int_0^\infty |u(r)|^2 dr = 1$$

One has shown earlier that radial wave function in region 11 is

$$u(r) = A \sin K(r - c)\ \ for\ c < r < b + c,$$

and the radial wave function in region 111 is

$$u(r) = Be^{-kr}\ \ for\ r > (b + c).$$

Hence, the total wave function over the entire region is calculated by integrating these functions over the region from c to $r > b+c$.

$$\int_0^\infty |u(r)|^2 \, dr = \int_c^{b+c} \left[A^2 \sin^2 K(r-c) \right] dr + \int_{b+c}^\infty B^2 e^{-2kr} \, dr \qquad (2.22)$$

This must be equal to 1

$$\text{or} \int_c^{b+c} A^2 \sin^2 K(r-c) dr = A^2 \int_c^{b+c} \frac{(1 - \cos 2K(r-c))}{2K} \, dr$$

$$A^2 \left[\frac{r - \sin 2K(r-c)}{2K} \right]_c^{b+c}$$

$$\frac{A^2}{2} \left[b - \frac{\sin 2Kb}{2K} \right] \qquad (2.23)$$

The integral

$$\int_{b+c}^\infty B^2 e^{-2kr} \, dr = B^2 \left[\frac{e^{-2kr}}{-2k} \right]_{b+c}^\infty = -\frac{B^2}{2k} e^{-2k(b+c)}$$

$$\frac{A^2}{2} \left[b - \frac{\sin 2Kb}{2K} \right] + \left[\frac{B^2}{2k} e^{-2k(b+c)} \right] = 1 \qquad (2.24)$$

where A and B are constants, whose values can be obtained by combining equation 2.24 with equations 2.14 and 2.15 as shown in the following:

$$A^2 = \frac{2k}{1 + kb}$$

$$B^2 = \frac{2k(\sin^2 Kb)}{(1 + kb)} e^{2k(b+c)}$$

2.7. Size of the Deuteron

One would like to compute the average of the square of the distance between the proton or neutron to the center of the mass of the n-p system.

r_d = separation between the n-p centers,

Taking the value of A^2 and B^2 from above and after partial integration of the above expression one gets

$$\langle r_d^2 \rangle = \frac{1}{8k^2} - \frac{1}{8K^2} + \frac{(2c+b)(1+kb)}{8k} + \frac{c^2}{4} - \frac{kb^3}{24(1+kb)} \qquad (2.25)$$

Proton is taken as point charge in the above equation.

$\langle r_d^2 \rangle$ is a function of c and b.

The expression $1/8K^2$ is relatively small for reasonable values of V_0.

Proton is not a point but has dimension, which is added to the above relation, and one gets

$$\langle r_d^2 \rangle = \frac{1}{8k^2} - \frac{1}{8K^2} + \frac{(2c+b)(1+kb)}{8k} + \frac{c^2}{4} - \frac{kb^3}{24(1+kb)} + \langle r_p^2 \rangle \qquad (2.26)$$

for the values assumed as

$V_0 = 40$ MeV, K= 0.955 F^{-1} and b = 1.895 F $k = 0.233 F^{-1}$, and $c = 0.4$ fm and $\langle r_p^2 \rangle = 0.8F$, one obtains a value for $\langle r_d^2 \rangle^{1/2}$ as equal to 2.22 fm. This is close to the value of deuteron radius experimentally determined by McIntyre et al. (2.4).

2.8. Tensor Forces

The magnetic dipole moment and electric quadrupole moment of the deuteron have been measured accurately. The quadrupole moment of deuteron as measured by Newell (2.5) and is $Q = 2.74 \pm 0.02 \times 10^{-27}$ cm^2. The existence of a quadrupole moment indicates that the nuclear force is not a purely central force since a central force in an S-state would give a zero value for the moment. In heavy nuclei, the quadrupole moments are caused by nuclear deformation, which is not the case for deuteron. The finite small value of deuteron's quadrupole moment indicates that the stable deuteron's bound state is a mixture of S and D states with a small contribution from D-state.

Moreover, in any nucleus, the magnetic moment is the sum of the dipole moments of the particles inside the nucleus. Deuteron consists of a proton and a neutron. Dipole moments of proton and neutron are as follows:

Proton dipole moment is = 2.79275 nm, neutron's dipole moment is = −1.91350 nm; therefore, the expected magnetic moment of deuteron is the sum of these two values as 0.87425 nm.

The measured value is 0.85735 nm, and there exists a discrepancy of 0.02190 nm or 2.5% from the expected value.

This discrepancy also indicates that deuteron is not in a pure $\ell = 0$ state, but the state is a mixture of different values of ℓ. The parity consideration will permit a mixture of $\ell = 0$ and $\ell = 2$ or a mixture of S-and-D wave configuration.

Total wave function of the deuteron for a mixture of $\ell = 0$ and $\ell = 2$ state can be written as

$$\Psi = a_S \, \psi_S + a_D \, \psi_D \qquad (2.27)$$

where ψ_s, ψ_D are normalized wave functions with $J^\pi = 1^+$ and the coefficients a_s and a_D satisfy the relation $a_s^2 + a_D^2 = 1$,

or one can say that deuteron is an S-state part of the time and in a D-state in part of the time. When it jumps from $S \rightarrow D$ state, the total spin has to flip over in order to conserve $(\ell + s) = J$. If the wave function ψ is not spherically symmetric, one obtains values of $a_D^2 = 0.04$ and $a_s^2 = 0.96$ in agreement with the observed value of deuteron's quadrupole moment.

Kellog et al. (Ke) measured the electric quadrupole moment of deuteron using molecular beam of deuterium gas and obtained a value of $Q_2 = 0.0023$ nm.

One does not expect a finite value of quadrupole moment for a spherical nucleus for the neutron and proton pair to be in $S = 0$ state. One obtains the correct value of the magnetic dipole moment if the deuteron is a mixture of S and D states, and contribution of D state is $a_D^2 = 3.9\%$.

2.9. Spin Dependence of Nuclear Forces

Two other nucleon-nucleon systems are (p-p) and (n-n) pair. First is known as diproton, and second is known as dineutron system.

According to Pauli's exclusion principle for identical particles, this pair can only exist in singlet states $S = 0$, whereas n-p pair in deuteron exists in $S = 1$ states. Since (n-n) and (p-p) pairs do not form bound states, it is fair to conclude that nucleon-nucleon force in a singlet state is weaker than in the triplet state, or the nuclear force is spin dependent.

2.10. Summary of the Nuclear Force

The information about the properties of nuclear force obtained from the deuteron problem is not complete. To obtain more information, one has to study behavior of nucleon-nucleon interaction and nuclei containing three or four nucleons. This will be discussed in the next chapter.

CHAPTER 3

NUCLEON-NUCLEON INTERACTION

3.1. Nucleon-nucleon Interaction

The most straightforward method of studying the interactions between particles is to perform scattering of nucleons with other nucleons.

Scattering of nucleons is produced as a result of nuclear force between them. The simplest of this study is that of 2-body problem that is scattering of neutrons by protons, protons by protons, and neutrons by neutrons.

The kinematics of n-p scattering reaction is shown below.

The relation between laboratory system and the center-of-mass system is shown in figure 3.1 below.

Assuming that $m_p = m_n$ approximately, then in the center of mass, the relation between scattering angles are given as

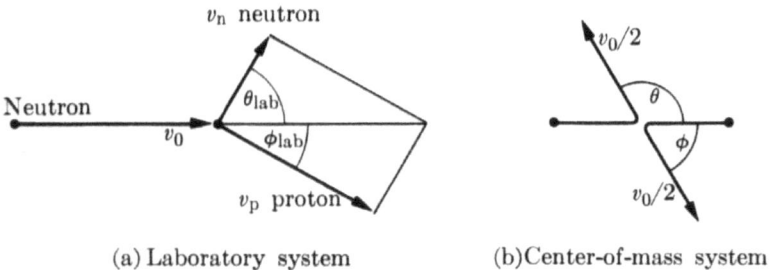

(a) Laboratory system   (b)Center-of-mass system

Figure 3.1 Relation of scattering angles and energies
in the laboratory and CM system.

$$\theta_{CM} = 2\theta_{Lab} \quad and \quad \phi_{CM} = 2\phi_{Lab}$$

and the energy of neutron in the center of mass is given as

$$E_{CM} = \frac{1}{2} E_{Lab}$$

Incident nucleons, neutrons, or protons are directed toward a target of n or p in the laboratory. However, theoretical calculations are made in the center of mass system.

3.2. Scattering Cross Section

The probability of scattering one particle by another is expressed in terms of a cross section, which is measured in unit of barns where $1b = 10^{24} \, cm^2$.

In actual measurements, this probability or intensity of scattered particles or cross section is measured by determining the number of particles scattered at a certain angle (θ) and dividing it by the number of particles incident or colliding with the target per unit second. In other words, the probability of scattering is given as

$$\text{Probability} = -\frac{dN}{N} = \frac{(nAdx)\sigma}{A} = n\sigma dx \qquad (3.1)$$

where $-dN$ is number of particles scattered by the target after interaction, N_0 is number of incident particles hitting the target nucleus, n = number of nuclei per unit volume of the target, and dx = thickness of the target.

Integrating 3.1 over the entire thickness of the target, one gets number of undeflected particles as

$$N = N_0 e^{-n\sigma x} \qquad (3.2)$$

and the number of particles scattered by the target is given as

$$N_{sc} = N_0(1 - e^{-n\sigma x}). \qquad (3.3)$$

If the scattered particles are measured at an angle (θ), then the total scattering cross section is obtained by summing the scattered particles over all angles, which is given as

$$\sigma = 2\pi \int_0^\pi \frac{d\sigma(\theta)}{d\Omega} \sin\theta \, d\theta \qquad (3.4)$$

where $\dfrac{d\sigma(\theta)}{d\Omega}$ is known as differential scattering cross section at an angle

(θ), and $(d\Omega)$ is the solid angle subtended by the detector.

3.3. Experimental Data on Low Energy Neutron-proton Scattering

Many scientists have measured the n-p scattering cross section over a wide range of neutron energies. We will confine our discussion to neutron energies in the range of 0.02-20 MeV (in the lab system).

The n-p scattering at very low neutron energies less than $0.1\ eV$ suffers from the effect of chemical binding of hydrogen molecules. These molecules are formed in two states: orthohydrogen with spins of two protons aligned parallel and parahydrogen with spins antiparallel. The scattering cross section from these two forms of hydrogen produces interference effects and gives rise to large scattering cross section.

Since the two nucleons cannot come closer than a certain distance, it is assumed that they experience a repulsive force for distance less than c known as hard core infinite repulsive potential from center to c as shown in figure 3.2.

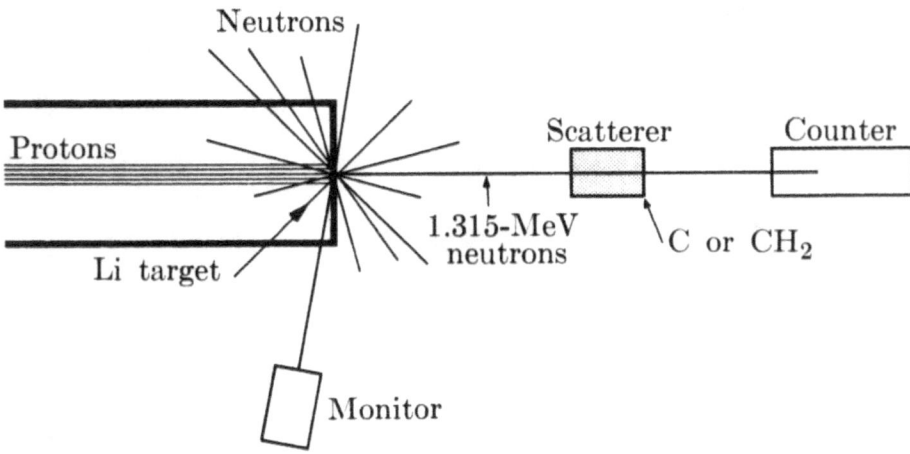

Figure 3.2. Neutron-proton scattering experiment by Storrs and Frisch

Mono-energetic neutrons of energy 1.315MeV are produced by the nuclear reaction Li^7(p, n)Be^7 by hitting a lithium target by protons accelerated in a Van de Graff accelerator (3.2). These neutrons are beamed through a collimator to a CH_2 gas target. Neutrons scattered from the target are measured at several angles and then summed over all angles to obtain value of total scattering cross section.

The experimental set up used by Storrs and Frisch (3.1) for measurements of n-p cross section above the neutron energy of about 200 keV is shown in figure 3.2.

Neutrons are scattered by both carbon and hydrogen nuclei. The scattering cross section due to carbon is subtracted from the results of each measurement to obtain cross section due to protons. Several measurements are made at different neutron energies, and the results of these measured cross sections compiled by R.K. Adair. are shown in the figure 3.3.

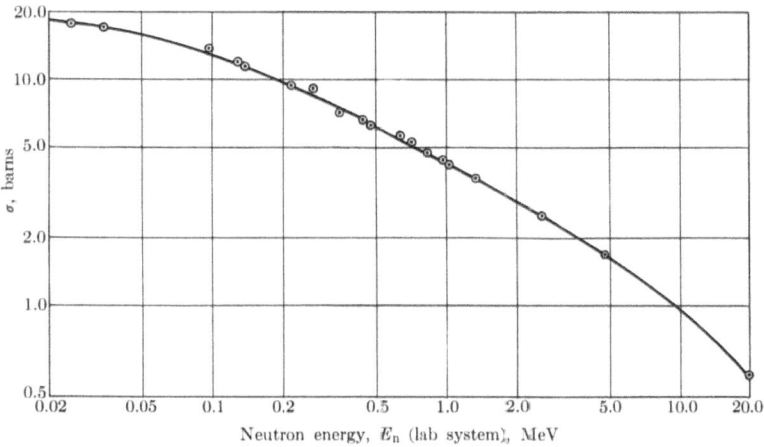

Figure 3.3. Total n-p scattering cross section in barns as a function of neutron energy

Theoretical curve is based on the parameters, $a_s = -23.7$ fm, $a_t = 5.38$ fm, $r_{0t} = 1.70$ fm, $r_{0s} = 2.4$ fm.

The experimental points are from a review article by R. K. Adair, *Review of Modern Physics* 22, (1950) p. 249.

At low neutron energies, the scattering cross section is predominantly due to *s*-wave neutrons.

3.4. Wave Mechanical Treatment of n-p Scattering

Similar to the wave mechanical treatment of deuteron problem, one can study the neutron-proton scattering using wave mechanics. Neutron is treated as a travelling wave with wave number $k = \dfrac{\sqrt{2mE_n}}{\hbar}$ incident on a proton at rest in the lab. E_n is the neutron energy, and m is the reduced mass given as $\dfrac{m_n m_p}{m_n + m_p}$. Wave function describing a beam of neutrons moving in the Z direction is

$$\psi = e^{-ikz} = e^{-ikr\cos\theta}$$

The neutron wave sees the potential of the proton and gets scattered by proton at different angles. The nuclear potential used in this analysis is similar to that used in the case of deuteron system except for the repulsive core, which is taken as $c = 0$.

The wave function describing a beam of particles of energy E and angular momenta (l) is expressed in a series in terms of spherical harmonic functions $Y_{l,0}(\theta)$ and radial wave function $u_l(r)$.

This series expansion describes particle waves of different momenta. The radial wave function can be expressed in terms of spherical Bessel function $j_e(kr)$ where k is related to particle energy, and r is the distance of the wave from the scattering center. Based upon the values of $j(kr)$ for values of $l = 0, 1, 2, 3, \dots$, it can be shown that at energies lower than 10 MeV, only particles with $l = 0$ will come close to the scattering center to be scattered.

Hence, for low energy neutrons, only s-wave scattering is dominant ($l = 0$) in the scattering since at low energies, higher angular momentum neutrons will not come close to the protons to suffer scattering.

Ignoring spin-orbit interaction and using a square-well nuclear potential, which is given as

$$V = -V_0 \ \text{for } r < r_0$$
$$V = 0 \ \text{for } r > r_0$$

One observes that when the force of interaction in the form of this potential is turned on as the wave approaches the proton center, nothing happens to the incoming part of the wave function, but the outgoing part of the wave suffers a phase shift $2\delta_0$ and a change in its amplitude.

Outside the range of nuclear potential, the outgoing wave does not incur any change in amplitude.

Resulting ($l = 0$) part of plane wave after scattering is given as

$$\psi_0 = \frac{e^{i(kr+2\delta_0)} - e^{-ikr}}{2ikr} = \frac{e^{i\delta_0}}{kr}\left[\frac{e^{i(kr+\delta_0)} - e^{-i(kr+\delta_0)}}{2i}\right]$$

$$\psi_0 = e^{i\delta_0}\frac{\sin(kr+\delta_0)}{kr}$$

(3.5)

This is a solution of $l = 0$ Schrödinger equation for a free particle (outside the range of nuclear potential).

New total wave function for scattering outside the scattering potential obtained by adding the original plane wave, one gets

$$\psi = e^{ikz} + \frac{e^{i(kr+2\delta_0)} - e^{-ikr}}{2ikr} = e^{ikz} + \frac{e^{i(kr+\delta_0)}}{kr} - \sin\delta_0$$

(3.6)

Outgoing scattered wave has an amplitude $\left(\dfrac{\sin\delta_0}{kr}\right)$ moving away form the scattering center.

Number of particles carried by this per second is found by integrating the flux over a sphere of radius r enclosing the scattering center, which is

$$N_{sc} = \left(\frac{\sin\delta_0}{kr}\right)^2 (4\pi r^2)v$$

(3.7)

where v is speed of incident neutron of energy E_n

$$N_{sc} = \left(\frac{4\pi\sin^2\delta_0}{k^2}\right)v$$

(3.8)

Incident flux of neutrons per second per unit area is v or the scattering cross section σ_{sc} is given as

$$\sigma = \frac{N_{sc}}{N} = \frac{4\pi\sin^2\delta_0}{k^2}$$

(3.9)

This shows that the total scattering cross section of neutrons by protons at low energies is related to neutron energy $k = \dfrac{\sqrt{2mE_n}}{\hbar}$ and phase shift δ_0 experienced by the wave of the scattered particle. It can be shown that σ_T at higher energies where other l values might be involved is given as

$$\sigma = \sum_l \frac{4\pi(2l+1)\sin^2\delta_l}{k^2} \tag{3.10}$$

Where δ_l is phase shift corresponding to angular momentum.

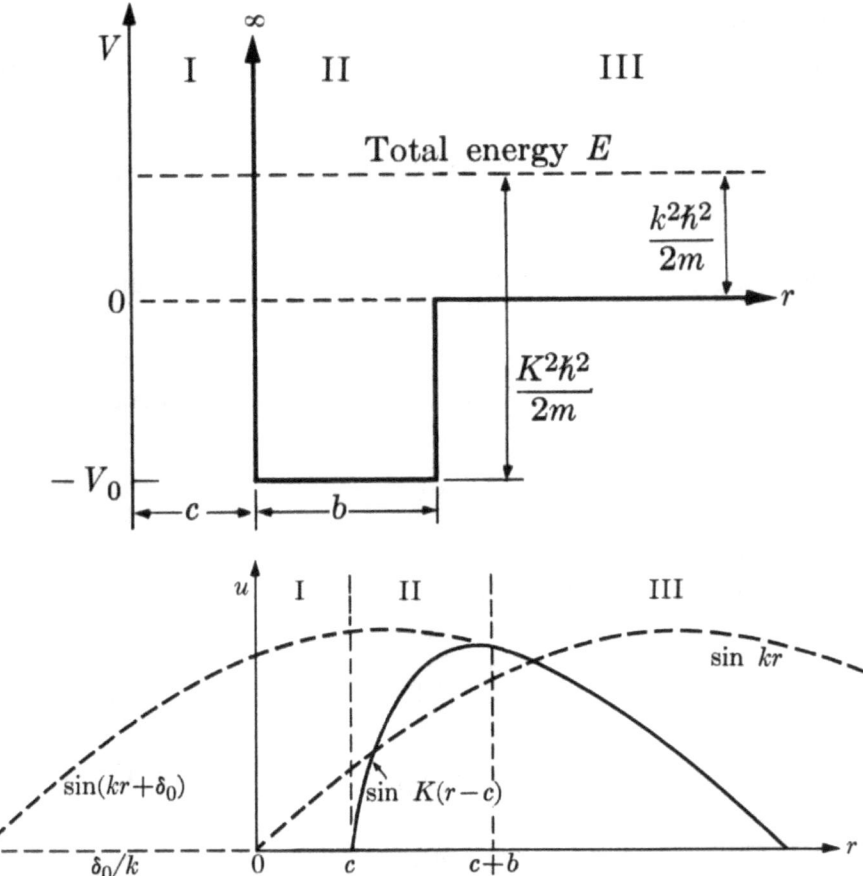

Figure 3.4. Square-well n-p potential and the radial function u for n-p scattering.

3.5. Determination of the Phase Shift δ_0

One would now determine the phase shift δ_0 for low energy n-p scattering by solving Schrödinger equation in the region of interaction.

We again assume that the potential is a scuare-well potential as

$$V = -V_0 \quad for \; r < r_0 \; and \; V = 0 \; for \; r > r_0 \tag{3.11}$$

In region II, the radial part of the wave function for $l = 0$ must satisfy the radial wave equation.

$$\frac{d^2u}{dr^2} + \frac{2\mu}{\hbar^2}(V_0 + E_n)u = 0 \tag{3.12}$$

$$u(r) = 0 \quad at \; r = c$$

where μ is the reduced mass. Solution of the above equation in the region II where $r < c + b$ is given as

$$u(r) = A\sin K(r - c) \tag{3.13}$$

Wave numbers K and k are given as

$$K^2 = \frac{2\mu(V_0 + E_n)}{\hbar^2} \quad and \; k = \frac{\sqrt{2\mu E_n}}{\hbar}$$

and the general solution for equation 3.12 in Region III where $V_0 = 0 \; r > (c + b)$ is given by

$$u(r) = B\sin(kr + \delta_0) \tag{3.14}$$

A and B are constants of normalization and δ_0-phase shift suffered by incident neutron.

Matching the wave functions and their derivatives at the boundary $r = c + b$, one gets the following equations:

$$A\sin(kr_0) = B\sin(kr_0 + \delta_0) \tag{3.15}$$

Differentiating equation 3.15 with respect to r, one gets

$$KA \cos Kr_o = kB \cos(kr_0 + \delta_0) \tag{3.16}$$

Dividing 3.16 by 3.14, one gets

$$K \cot Kr_0 = k \cot(kr_0 + \delta_0) \tag{3.17}$$

where $r_0 = c + b$.

From the study of deuteron problem, we had earlier obtained values of $V_0 = 73 Mev$ and $r_0 = 1.337\, fm$ for the depth of the potential and its range respectively.

Using these values in the above expression 3.17, one can obtain values of δ_0 as a function of neutron energy E_n. Knowing the value of δ_0 and k, one can obtain σ_0 from the relation

$$\sigma_0 = \frac{4\pi}{k^2} \sin^2 \delta_0. \tag{3.18}$$

The calculation of σ_0 versus E_n based upon the nuclear potential and its range obtained from the triplet case is compared with experimental data, which is shown in the figure 3.5.

The agreement is good for $E > 5$ MeV but deviates significantly at energies < 5 MeV. The discrepancy is due to the fact that deuteron was a triplet state ($3S_1$, J = 1) formed with parallel spins whereas n-p scattering takes place in both singlet $S = 0$ and triplet $S = 1$ states.

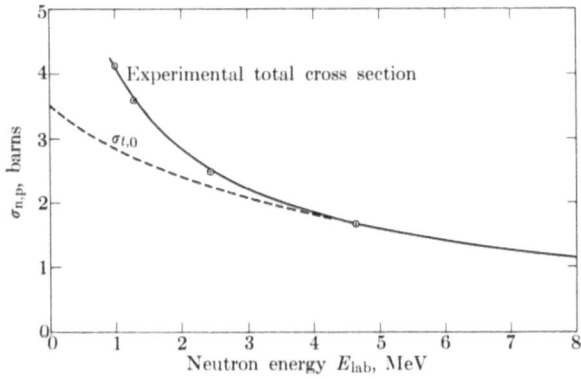

Figure 3.5. Measured and calculated n-p
scattering cross section versus neutron energy

3.6. Relation of δ_0 and Scattering Length (a)

A quantity (a) known as the scattering length is defined as the intercept of the radial wave function with the r-axis outside the range of nuclear force (asymptotic solution).

In the measurement of n-p cross section, we obtain values of phase shift δ_0, which is related to scattering cross section. There is also a relationship between phase shift δ_0 and scattering length (a) if the potential well (V) is deep more than 50 MeV, and the range is short (r_0) so that the radial wave function $u(r)$ at the boundary $r = b + c$ has a negative slope, and the straight line of the slope would intersect the r-axis at a positive distance a_1 from the origin.

The equation for $u(r)$ outside the range of nuclear force is a straight line intersecting the r-axis at $r = a_1$. This is shown in figure 3.6.

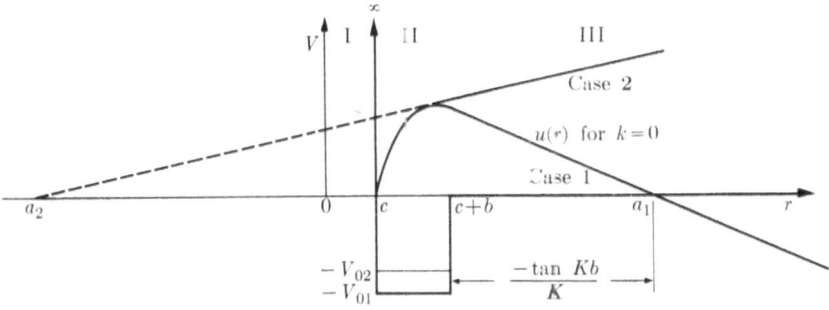

Figure 3.6. Radial wave function u in the limit of zero neutron energy. Case 1 (potential-V_{01} binding) gives positive scattering length; Case 2, negative (potential-V_{02} nonbinding).

3.7. Scattering Lengths for Bound and Unbound System

The following cases are discussed below.

Case 1. This situation is for the bound state of (n-p system)

The radial wave function outside the range of nuclear potential in region III in figure 3.6 is given as $u(r) = B\sin(kr + \delta_0)$ at $u(r) = 0$ or at $r = a_1$ and, hence,

$$B \sin(ka_1 + \delta_0) = 0, or \ \sin(ka_1 + \delta_0) = 0$$
$$or \ ka_1 + \delta_0 = \pi$$
$$or \ \delta_0 = \pi - ka_1$$

This gives the relation between δ_0 and scattering length a_1 for a bound state.

Case 2. This applies to the unbound state of n-p system

If the nuclear potential is not deep enough as is the case for the (n, n) and (p, p) pairs, it will not form a bound system. Then the slope at the boundary is positive, and it would intercept the r-axis at a negative value of $-a_2$ (as shown in figure3.6).

Scattering length a_2 therefore is given as $\sin(ka_2 + \delta_0) = 0$ and

$$ka_2 + \delta_0 = 0, \text{ or } \delta_0 = -ka_2).$$

It has been shown that n-p scattering cross section is given as

$$\sigma_0 = \frac{4\pi \sin^2 \delta_0}{k^2} \tag{3.19}$$

Substituting the value of phase shift and the scattering length in the equation 3.19, one gets the value of scattering cross section as $\sigma_0 = 4\pi a_1^2$ for bound system, which is a triplet state (J = 1) with parallel spins and $\sigma_0 = 4\pi a_2^2$ for unbound singlet state (J = 0) with antiparallel spins.

3.8. Singlet and Triplet Potentials

The n-p scattering cross section takes place with both parallel and antiparallel spins forming $S = 0$ and $S = 1$ states. These cross sections are incoherent and therefore add algebraicly. Hence, the n-p scattering cross section is the sum of scattering in both states. Since triplet state has three substates, and a singlet state has one substate, the total cross section can be written as

$$\sigma = \frac{3}{4}\sigma_t + \frac{1}{4}\sigma_s \tag{3.20}$$

where σ_t and σ_s are the total cross sections for the triplet and singlet states.

The measured value of σ_{np} cross section at low energy is 20 barns as shown in figure 3.3. Hence, by putting the value in 3.20, one gets

$$20 = \frac{3}{4}\sigma_t + \frac{1}{4}\sigma_s$$

where σ_t value is between 3 to 5 barns as calculated using values of $V_0 = 73.3$ MeV and $r_0 = 1.33$ fm for the strength and range of the nuclear potential obtained from deuteron problem in ^3S state. The actual value is 4.0 barns. Using the relation, 4 barns = $4\pi\,(a_t)^2$, gives triple scattering length value as $a_t = 5.38$ fm. The scattering length is positive for a bound system.

Using the value of $\sigma_s = 68$b, one obtains a value of singlet scattering length $a_s = -23.2$ fm. For an unbound system, the scattering length is negative.

Hence, values of singlet and triplet scattering lengths are respectively as follows:

$a_s = -23.2$ fm and $a_t = 5.38$ fm.

Thus, from n-p scattering cross-section measurement, one obtained values for both the triplet bound state and the singlet unbound state. These values would be compared with similar values obtained from p-p and n-n scattering cross sections.

3.9. The p-p Scattering Cross-Section Measurements

The p-p scattering cross section differs in some respects with n-p scattering. First is that since both particles are positively charged particles, they produce Coulomb repulsion, which for low energies of protons, will not permit the two particles to come close as is known from the fusion reaction in the sun. At energy less than 100 keV, there is zero probability of interaction. The probability increases slowly as the proton energy increases. For this reason, most measurements of p-p scattering are done for energies greater than about 1 MeV.

Second, in view of the fact that two protons are identical, it would be impossible to identify which proton is which—incident or the target proton. According to Pauli's exclusion principle, the identical particles cannot have the same quantum numbers; therefore, the p-p scattering will take place with their spins antiparallel, that is, in $S = 0$, the singlet state. Therefore, study of p-p scattering cross section will only provide information about the nuclear force in state with antiparallel spins.

Third, the p-p scattering cross section will be dominated by the Coulomb scattering in the forward direction, and the cross section will be anisotropic. Therefore, one measures the differential scattering cross section as a function of angle, and after subtracting the Coulomb scattering, one determines the nuclear scattering cross section.

It is, however, easy to study p-p interaction. Since proton beams of several MeV with great accuracy of energy can be easily made available in particle accelerators such as a Van de Graaff accelerator, and proton targets are also easily available in the form of hydrogen gas at high pressure.

Extensive measurements have been made for many proton energies, extending up to few 100 MeV.

At low proton energies, only s-wave interaction takes place. Whereas at higher proton energies, angular momentum l>1 are involved. Measurements are performed at fixed energies of protons. Faraday cup collects the proton beam and measures the charge, and hence, the incident beam intensity. Protons that are scattered at an angle θ are collimated and are detected by a proton detector. The detector assembly can be rotated about the center of the target chamber, and θ can be varied and $\left(\dfrac{d\sigma}{d\Omega} \right)$ can be measured at given angle θ; $d\Omega$ is the solid angle subtended by detector. When the incident beam is unpolarized, i.e., spin direction are randomly distributed, $\left(\dfrac{d\sigma}{d\Omega} \right)$ is independent of φ and

$$\left(\frac{d\sigma}{d\Omega} \right)_\theta = \frac{N_{sc}(\theta)}{N_{indicent}}$$

Potential energy diagram for p-p case, which consists of nuclear and Coulomb potential is shown in figure 3.7.

Nuclear potential is attractive with a chosen value of $V_0 = -40$ MeV. The Coulomb potential is repulsive. Nuclear potential is much greater than the Coulomb potential. One can see this behavior in figure 3.7.

Radial wave equation for proton energy $E_p < 5\,Mev$ and for $l = 0$ is

$$\frac{d^2u}{dr^2} + \frac{2\mu}{\hbar^2}\left[E_p - V_N(r) - \frac{ke^2}{r}\right]u = 0 \qquad (3.21)$$

where $V_N(r)$ is square-well potential as has been discussed earlier for the n-p case, ke^2/r is the Coulomb potential, and E is the proton energy in the center of mass.

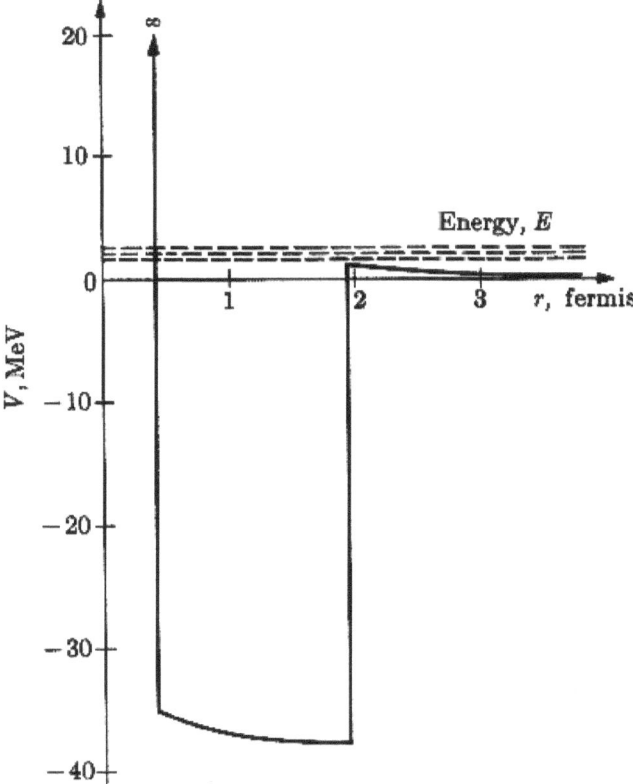

Figure (3.7)   Nuclear and Coulomb potential for p-p scattering.

Similar to n-p case, the accepted solution of the equation 3.21 outside the range of nuclear potential is

$$u(r) = Sin(kr - \eta \ln 2kr + \sigma_0 + \delta_{0)})$$

where σ_0 is the Coulomb phase shift and δ_0 is the nuclear phase shift, $\eta \ln 2kr$ is due to infinite range of the Coulomb force where $\eta = \dfrac{ke^2}{\hbar v}$.

The scattering formula for low energy p-p differential scattering cross section $\dfrac{d\sigma}{d\Omega}$ at a given proton energy is given as

$$\frac{d\sigma}{d\Omega}(\theta) = \frac{\eta^2}{4k^2}\left[\frac{1}{\sin^4\frac{\theta}{2}} + \frac{1}{\cos^4\frac{\theta}{2}} - \frac{\cos(\eta\ln\tan^2\frac{\theta}{2})}{\sin^2\frac{\theta}{2}\cos^2\frac{\theta}{2}}\right] + \Delta_{vp}$$

$$-\eta\frac{\sin\delta_0}{2k^2}\left[\frac{\cos(\delta_0 + \eta\ln\sin^2\frac{\theta}{2})}{\sin^2\frac{\theta}{2}} + \frac{\cos(\delta_0 + \eta\ln\cos^2\frac{\theta}{2})}{\cos^2\frac{\theta}{2}}\right] + \frac{\sin^2\delta_0}{k^2}$$

(3.22)

Equation 3.22 contains for l = 0 only one adjustable parameter that is S-wave phase shift δ_0 since the Coulomb potential is known, and the Coulomb scattering cross section can be calculated. For a given proton energy, one can calculate k as

$$k = \frac{\sqrt{2\mu E_p}}{\hbar} \text{ and } \eta = \frac{e^2}{4\pi\varepsilon_0\hbar v}$$

(3.23)

where v is the relative velocity of the two particles, Δ_{vp} is a correction term.

Experimental measurements of p-p differential scattering cross section as a function of scattering angle θ_{CM} carried out by Knecht et al. (3.4) for three proton energies E_p = 1.397 MeV, E_p = 1.855 MeV and E_p = 2.42 MeV is shown in figure 3.8.

From the analysis of these experimental measurements, one obtains values of scattering phase shifts as shown in the figure at three energies of protons respectively.

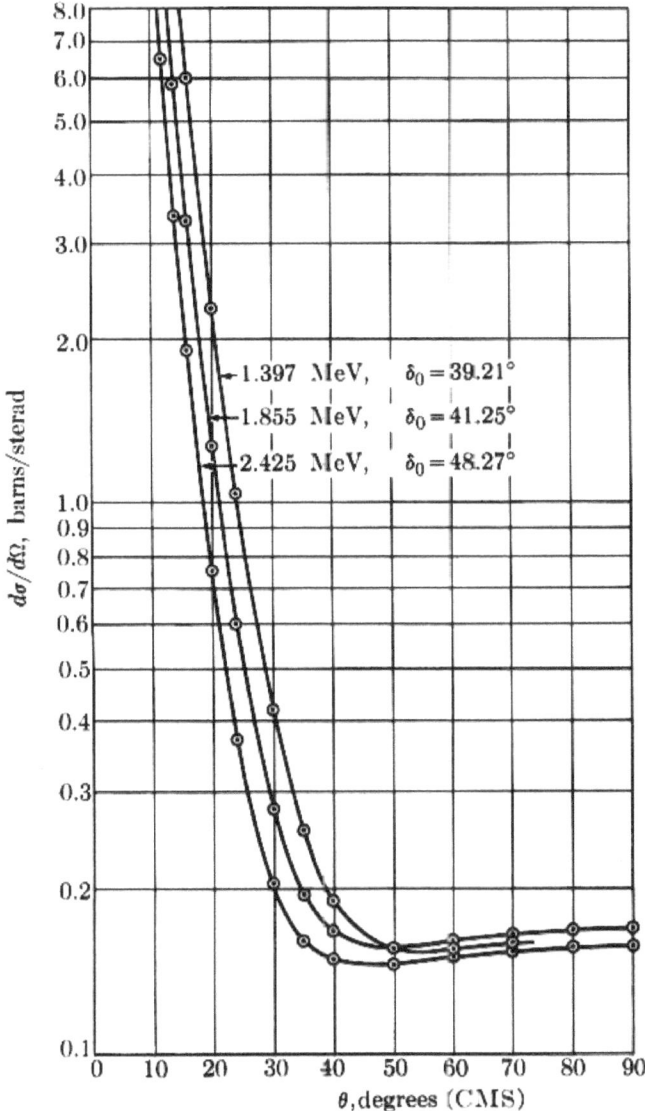

Figure 3.8. Theoretical and experimental p-p cross sections at three energies (from data given by D. J. Knecht et al.)

Observed scattering cross section for p-p interaction is a complicated function of proton energy, the scattering angle θ and the phase shift $-\delta$. At small angles up to $\theta = 30$ degrees, the cross section is predominantly due to Coulomb force, and in the middle range of angles scattering cross section is isotropic, and this is mainly due to the nuclear force. One can determine the nuclear potential parameters (V_0, b) from the knowledge of δ_0 obtained from the measurement as follows:

$$K \cot Kb = C^2 k \cot \delta_0 + D^{-1} h(\eta) \qquad (3.24)$$

where $C^2 = \left[\dfrac{2\pi\eta}{\exp(2\pi\eta) - 1} \right]^2$ is the Coulomb penetration factor, $D = 2.88 x 10^1$

cm is the proton radius characteristic of Coulomb scattering, and $h(\eta)$ is a slowly varying function of proton energy.

From the determination of the phase shift δ_0 for $l = 0$, one determines the scattering length value obtained from p-p cross section. Since due to Pauli exclusion principle, the p-p scattering takes place only in a singlet state and since diproton system is unbound, the measured scattering length will have a negative sign. Within the range of nuclear force, the Coulomb force can be treated as a small perturbation. Thus, eliminating the contribution from the Coulomb scattering, one can determine the nuclear scattering length, which can be given as

$$(a_p)^{-1} = a_n^{-1} + D(\ln D / r_0 + 0.33) \qquad (3.25)$$

where a_n is the neutron-proton scattering length for same nuclear potential and r_0 is the range of nuclear force. Constant 0.33 is an estimate based upon the shape of the nuclear well.

Substituting the value of a_n and r_0 determined from the experiment, one obtains a value for $(a_p)^{-1} = -17.0 x 10^{-13}$ cm,

which is comparable to the value of $a_n = -24.3$ fm obtained from the n-p scattering cross section.

3.10. Study of n-n Scattering Cross Section

Since neutron is an unstable particle, one cannot have a target of neutrons. To study n-n interaction, one studies n-d scattering cross section and then subtracts the contribution of scattering from protons.

These results have large experimental errors. However, from these studies, one measures the phase shifts (δ_0) of the singlet ($J = 0$) scattering length a_s of the unbound state. The n-n scattering length deduced from 3-body problem is

$$a_{nn} = -17.06 \pm 1.00 \text{ fm}$$

3.11. Effective Range Theory

Theoretical calculation of n-p cross section in the neutron energy range of 0-20 MeV have been made based upon the effective range theory. We know that n-p scattering cross section is due to both singlet and triplet states.

Using the radial wave equations for two energies outside the range of nuclear potential allows one to obtain a relation between phase shifts δ and effective range of potential in both states. The relation is given as

$$\cot \delta_0 = -\frac{1}{ka} + \frac{1}{2} kr_o \text{ and}$$

$$\sigma = \frac{1}{4} \left[3 \frac{4\pi a_t^2}{a_t^2 + k^2 + \left(1 - \frac{1}{2} a_t^2 kr_{0t}^2\right)^2} + \frac{4\pi a_s^2}{a_s^2 k^2 + (1 - \frac{1}{2} a_s k^2 r_{os}^2)^2} \right] \qquad (3.26)$$

Using values of a_t and r_{0t} obtained from the triplet state of deuteron as $a_t = 5.38$ fm, $r_{0t} = 1.70$ fm, one obtains values of a_s and r_{os}. The results obtained from experiments of n-n, p-p, and n-n data have provided values of singlet and triplet scattering lengths and effective ranges of the nuclear potential. These are given in table 3.1.

Table 3.1. Scattering lengths and range of the nuclear potentials for the nucleon pairs *n-p*, n-n, and p-p

Isotopic spin	spin	type	a (fermis)	r (fermis)
1	0	n-p	-23.74 ± 0.09	$r_s = 2.2F$
1	0	p-p	-17.00	$r_s = 2.4$ F
0	1	n-p	5.39 ± 0.03	$r_t = 1.7$ F
0	0	n-n	20 ± 2.6	

The difference between n-p and p-p scattering lengths is partly due to Coulomb effects.

3.12. Nucleon-nucleon Force at High Energies

So far, our discussion of the nucleon-nucleon force had been obtained from scattering cross-section measurements at low energies below 20 MeV where only *s*-wave scattering plays a dominant role, however, at higher energies scattering with l > 0 becomes significant and the cross-section measurements provide information about phase shifts δ_l for higher angular momenta.

Extensive measurements of n-p and p-p measurements have been made by G. Breit et al. (3.5) at Yale University in energies between 100 and 300 MeV.

Figure 3.9 shows results for n-p differential scattering cross-section measurement for neutron energies at 215, 260, and 300 MeV by Hull et al. (3.6).

These measurements show that n-p scattering cross section is anisotropic, that is, neutrons are scattered in the forward as well as in backward directions showing significant contribution from l > 0 scattering. The scattering phase shift as well as total cross section decreases monotonically as the neutron energy increases.

Differential scattering cross section of high energy neutrons with protons in the center of mass due to Yang and Wolfenstein (3.7) is expressed as

$$\frac{d\sigma(\theta)}{d\Omega} = \sum_{l=0}^{2l\,max} A_l Y_{l,s}(\theta)$$

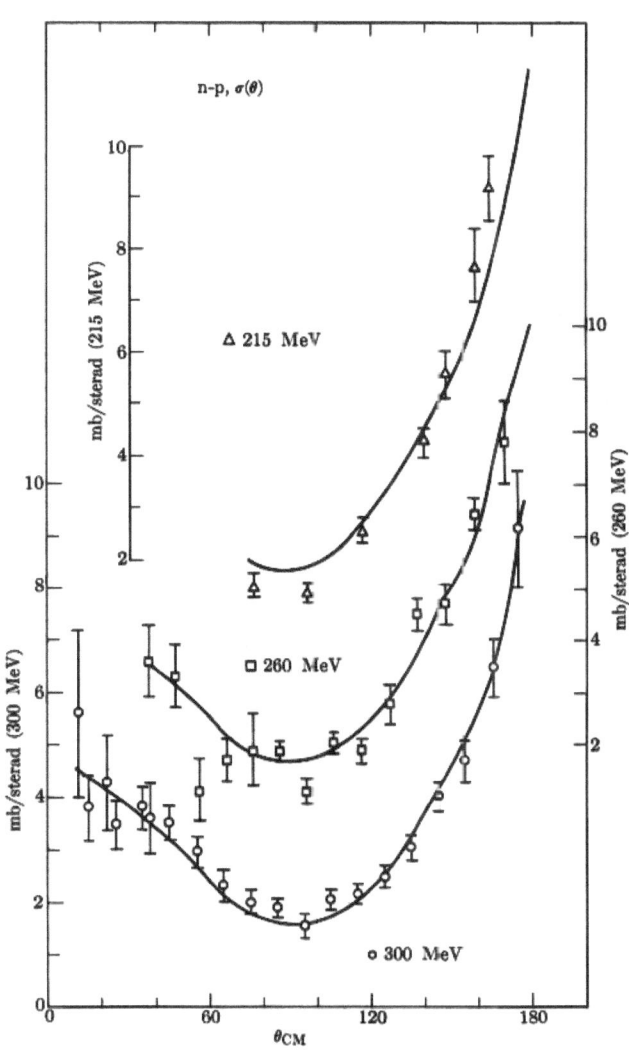

Figure 3.9. The n-p differential cross section at 215, 260, and 300 MeV measured and calculated from Breit's phase parameters (from M. H. Hull)

Summation is over $(2l + 1)$ terms—each term for each value of l from 0 to maximum value of l determined by the particle energy. The theoretical calculation involves determination of phase parameters for each value of l and mixing parameters for different l-values contributing to the cross section.

In p-p scattering, Pauli exclusion principle applies, which implies that $S = 1$ scattering can occur in odd parity states, and $S = 0$ scattering takes place in even parity states, thus, restricting the behavior of cross section to odd or even angular momenta. The p-p experimental data shows a forward peaking due to Coulomb repulsive force and an isotropic cross section at larger angles due to nuclear scattering.

The analysis of experimental data requires calculation of phase shifts for many angular momenta and is rather complicated. The agreement with the experimental data requires use of exchange potentials. Serber type of exchange potential has gained success in fitting the data.

3.13. Polarization Experiment

Angular distribution measurements alone does not yield enough information to determine the phase shifts (δ_l) and mixing parameters of different l-values at the given energy. Additional information can be obtained from polarizing experiments, i.e., measurement of spin orientation of scattered particles—such experiments involve double scattering.

3.14. Summary of Nuclear Force

In summary, the properties of the nuclear force as obtained from the nucleon-nucleon scattering measurements can be summarized as follows:

Nuclear force is attractive and has short range. At very short distance, the force becomes repulsive.

Nuclear force is spin dependent. Singlet force is weaker than the triplet force.

Nuclear force is charge independent, that is, the force between n-n, p-p, and n-p pairs are the same. However, high energy results put some doubt on this statement.

Nuclear force depends on the relative orientation of spin and displacement vector (s.r.). This is known as a tensor force.

Nuclear force is also dependent upon particle energies or on the l-value (obtained from high energy nucleon-nucleon scattering experiments).

3.15. Three and Four Nucleon Systems

In addition to two-body nuclear force discussed above, one would also like to know if the two-body nuclear force is affected by the presence of other nucleons in the close proximity. For this, one can study the properties of He^3 and H^3 nuclei. Both of these nuclei have three nucleons. The binding energy per nucleon pair of H^3 is -2,827 MeV whereas the binding energy per nucleon pair for He^3 is -2.573 MeV. The He^3 is a stable nucleus whereas H^3 is unstable and decays by β-emission to He^3. Their radii are 1.70 and 1.87 fm respectively. We know that the radius of deuteron is about 2.2 fm, which means that H^3 and He^3 are more tightly bound than deuteron. The measured value of magnetic moments of these nuclei suggest a mixture of S and D states with a small contribution from the D state as is the case for deuteron. The nucleus He^4 is the only four-nucleon stable system since H^4 and Be^4 do not exist in nature. Properties of three-and-four nuclear system have been studied in great detail, and one may conclude that the force between two nucleon system is not affected by the presence of other nucleons in their vicinity.

3.16. Exchange Forces

The first explanation for the nucleon-nucleon force was given by Yukawa. According to him, this force was due to an exchange of a pi-meson between the two nucleons. This exchange of a virtual meson gives rise to an exchange force. Different exchange forces have been also proposed for this force, which depend upon the quantum properties of nucleons. For example, Majorana exchange force known as space-exchange force, Bartlett force known as spin-exchange force, and Heisenberg force known as space-spin exchange force.

Many scientists, following the example of Yukawa, have suggested that mesons heavier than pi meson are involved in producing nucleon-nucleon force at shorter distances. Some of these mesons are known as rho and omega mesons. The detailed discussion of exchange forces is beyond the scope of this book.

3.17. Elementary Particles and Quarks

We have discussed the nature of nuclear force between nucleons, protons, and neutrons. This discussion is based on the properties of deuterons and on the scattering cross-section measurements of n-p, p-p, and n-n reactions.

Before 1950, elementary particles known were protons, neutrons, and electrons, and some mesons observed in cosmic rays. After the development of high energy particle accelerators, one observed creation of many new particles, which are named hadrons and mesons. By 1960, scientists began to look for an answer to such a large number of elementary particles with specific mass, charge, and other quantum properties. In 1962, Gell-Mann (3.8) proposed an arrangement of certain particles in groups of eight and ten particles. He characterized this order, borrowed from the teachings of Buddha, as the "eightfold way." This scheme was so successful that he was able to predict the existence of an unknown particle called omega particle. This particle was later discovered at Brookhaven National Laboratory located in Long Island, New York.

Scientists wondered whether these known and newly discovered particles were indeed elementary, or they were made up of other elementary particles. In 1964, Gell-Mann (3.9) and Zweig proposed independently that these baryons and hadrons were made up of three particles. Gell-Mann named these particles as quarks. In order to satisfy the charge of baryons, he had to assume these quarks had fractional charges such as 1/3 or 2/3 units of charge; some had positive and some negative charge. For example, a proton, which has one unit of positive charge is made up of 3-quark (uud) combination and a neutron, which has no net charge is made up of 3 quarks (udd). The u quark had +2/3 units and d quark has -1/3 units of charge. Many scientists at the time did not accept this concept of an elementary particle with a fractional charge.

A series of measurement made by scientists with Stanford linear accelerator striking a target of protons with very high energy electrons showed that protons had three separate location of charges giving validity to this theory. Later on, more quarks were discovered, and now there are six quarks and six antiquarks. The field of elementary particle physics have been pursued vigorously in the last fifty years or so by thousands of scientists working in research centers all around the world. These concepts have produced new theories about the nucleon-nucleon force. Now the nuclear force between nucleons is that between quarks. This force between quarks is produced by an exchange of a massless particle called gluon. List of quarks and their charges, spins, and masses are given in table 2.1 in chapter 2.

CHAPTER 4

NUCLEAR STRUCTURE

4.1. Nuclear Structure and Nuclear Model

Atomic structure is universally based upon a nucleus at the center and electrons located outside the nucleus and revolving around it under the influence of electromagnetic force. Electrons are supposed to have quantum properties such as angular momentum, spin momentum, etc. These properties dictate the distribution of electrons in atomic shells with number of electrons in each shell determined by the quantum numbers.

Hence, it was natural to assume that neutrons and protons inside the nucleus be also distributed in a similar fashion dictated by the quantum properties of particles. One distinguishing feature of the nucleus is that it contains two types of particles—charged particle, proton and a neutral particle, neutron—whereas outer parts of atom have only electrons.

Previously, it has been shown that nuclear force is charge independent, i.e., the nuclear force between a pair of protons is the same as between a pair of neutrons. The presence of protons does not affect the nuclear force between pair of neutrons. Therefore, it was natural to assume that the distributions of protons inside the nucleus is not affected by the distribution of neutrons.

In the shell model of the nucleus, both neutrons and protons are treated as independent particles.

It will become apparent that the shell model proposed here has limited validity applicable only for low mass nuclei. The properties of medium and heavy mass nuclei do not lend support for this model, and other nuclear models have been proposed for such heavier nuclei.

Nuclear models that have been proposed over the years by scientists are known as follows:

1. Shell model or also known as independent particle model.
2. Collective model is based upon the deformation of nuclei.
3. Liquid drop model was used to explain the nuclear fission.
4. Fermi gas model was used to describe the level densities of nuclei.

We will now discuss the main features of each of these models and their successes and failures.

4.2. Shell Model or Independent-particle Model of the Nucleus

In this model, neutrons and protons are assumed to be distributed in shells; each shell carries certain number of neutrons or protons dictated by their quantum numbers such as angular momentum (l), spin quantum number (s), and magnetic quantum number (m_l).

All these nucleons see a central spherically symmetric potential produced by all the nucleons of the nucleus.

Motion of particles under this potential is such that the motion occurs in pairs of particles, i.e., for two protons or two neutrons in a given shell. If one particle has spin $\frac{1}{2}$, the other must have spin $(-\frac{1}{2})$, or if one particle is spinning in clockwise direction, the other particle spins in anticlockwise direction. The properties of the nucleus is thus determined by the odd or unpaired nucleon whose motion is determined by the average spherically symmetric potential produced by all the nucleons.

In determining the nature of nuclear force or potential, it was found that the shape of the potential could be a simple square well with finite or infinite depths or a shape with rounded edge known as Wood-Saxon potential.

For our discussion, we will consider two forms of potential:

1. Square-well potential with finite or infinite depth
2. Harmonic oscillator potential well

4.3. Natural Abundance of Elements Found on Earth

Historically, the study of atomic masses has been of the greatest importance in the development of atomic physics. As early as 1910, Soddy conjectured that some elements consisted of mixtures of atoms of different masses but similar chemical properties. These are known as isotopes.

Important progress came in 1919 with the development by Aston (4.1) of a mass spectrograph. Mass spectroscopy continues to play an important role in nuclear physics.

4.4. Relative Abundance of Atomic Species

The table of elements of stable nuclei show interesting features. For example, even number of protons (Z) and even number of neutrons (N) are strongly favored whereas nuclei with odd Z and odd N are least favored. The table shows atomic abundances of elements as was in 1950.

Table 4.1- Abundance of nuclei

Z	N	# of nuclei found in nature
Even	Even	166
Even	Odd	56
Odd	Even	56
Odd	Odd	9

Nucleosynthesis of Elements

Our planet has many elements from the lightest element of hydrogen to heavy transuranic elements such as thorium, uranium, etc. These elements are found on our planet in different abundances; some elements like hydrogen and helium are found in great abundances; and rare earth elements are very rare.

According to the theory of the creation of the universe, the universe was created 15 billion years ago by the explosion of enormous amount of energy called big bang. According to Einstein's famous discovery, E = MC^2, the energy converted into mass by creating atoms of the lightest element, hydrogen. Heavier atoms were produced by nuclear synthesis and successive capture of protons and neutrons by lighter elements to form heavier elements.

The synthesis of these elements was brought about by nuclear reaction at very high temperatures of billion degrees inside the core of high-mass stars. After the death of these stars, these elements were blown out from the decaying stars in the interstellar medium. These elements were then captured by new evolving stars like our sun. Our planet Earth was subsequently formed along with other planets in the solar system. For nuclei with certain number of protons and neutrons, the capture of neutrons or protons was less favored than by others making those nuclei more abundant than others.

Nuclear physicists investigated the reason for such abundances of elements in our planet. Nuclei with proton number Z = 2, 8, 20, 28, 50, 82, and nuclei with neutron number N = 2, 8, 20, 28, 50, 82, 126 were found to be more abundant than others. These nuclei were called magic nuclei, and these numbers were called magic numbers.

Shell model of the nucleus was proposed to explain the existence of magic numbers.

4.5. History of the Development of the Shell Model

1. Pre-1935 - The model was suggested as a possible explanation of the fluctuations in the relative abundance of nuclei in the periodic table. Such fluctuations were associated with the shell fillings and shell closures at magic numbers of fixed number of neutrons and protons.
2. 1935-1945 - Model failed to fit the binding energies of light nuclei and known magic numbers, hence, it was abandoned.
3. 1945-50 - This was the critical period for the survival of shell model.

In 1949, Fermi (4.2) made the suggestion that Pauli principle could strongly inhibit collisions between nucleons in certain orbits, and Mayer, Haxel, Jensen, and Suess (4.3) independently proposed the modification to the average potential by introducing a spin-orbit interaction ($V_{l \pm s}$), which gave the correct intrinsic spins of nuclei, and it predicted correct values of magic numbers as well as correct values of magnetic dipole moments.

4.6. Shell Model: Basic Properties

In a nucleus with many nucleons, it is almost impossible to derive an exact calculation of nuclear properties using the two-body nucleon-nucleon force. One can use only approximate methods.

First approximation used is that the force on any nucleon inside the nucleus is represented by a potential well due to all other nucleons. This is known as shell theory potential. It is also known as an independent particle model potential. This model is closely related to the case of atomic structure, that is, the nucleons are moving in fixed orbits under the influence of a central force just like the electrons in an atom moving around a nucleus under the influence of the Coulomb force.

Thus, IPM (independent particle model) provides a highly useful approximation to the total wave function of the nucleus.

4.7. Square-well Potential

Earlier, the square-well potential was used for the analysis of deuterons to determine the strength and range of the nuclear force between the pair of a neutron and a proton, which provided a value of about 73 MeV for the strength and 1.33 fermis for the range of the nuclear potential.

For the shell model, we use a similar square-well potential given as

$$V = -V_0 \quad r < R$$
$$V = 0 \quad r > R$$

where R is the nuclear radius. Schrödinger's equation of a single nucleon in this spherically symmetric potential can be given as

$$\psi = R_l(r) Y_{l,m}(\theta, \phi) \tag{4.1}$$

and the radial wave function $R_l(r)$ for any value of angular momentum l is given as

$$\frac{1}{r^2}\frac{d}{dr}(r^2\frac{dR_e}{dr})+\frac{2m}{\hbar^2}\left[E-V_0-\frac{l(l+1)\hbar^2}{2mr^2}\right]R_l=0 \qquad (4.2)$$

where E is the total energy of the nucleon in a given orbit and m is the reduced mass, which, in a heavy nucleus, is equal to the nucleon mass. The solution of equation 4.2 are the spherical Bessel functions, which are given as

$$R_l(r)=j_l(kr)=-\left(\frac{r}{k}\right)^l\left(\frac{1}{r}\frac{d}{dr}\right)^l\frac{\sin kr}{kr} \qquad (4.3)$$

where $k=\dfrac{\sqrt{2m(E-V_0)}}{\hbar}$ \qquad (4.4)

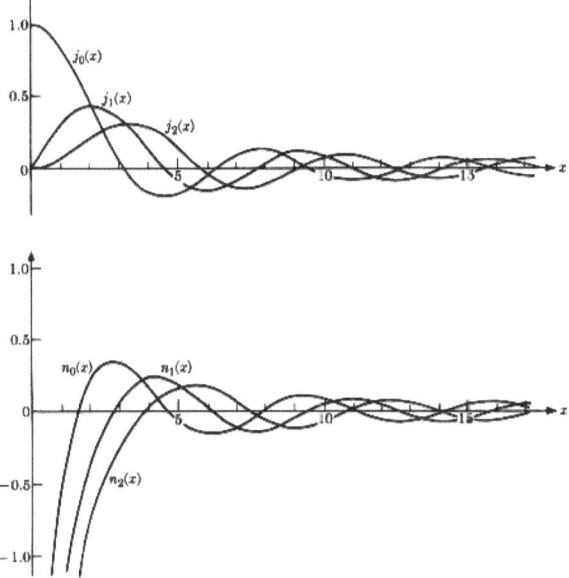

Figure 4.1. Spherical Bessel and Neumann function for $l = 0$, 1, 2 (From R. H. Dicke and J. P. Wittke, *Introduction to Quantum Mechanics*. Addison Wesley, 1960)

The three lowest order spherical Bessel functions for $l = 0, 1, 2$ are plotted in figure 4.1, where $x = kr$.

Another boundary condition required for the Schrödinger equation 4.1 is that the wave function be zero for all values of θ, and z at $r = r_0$ whose value is given in fermi.

Any value of $x = kr$ that makes a spherical Bessel function equal to zero is denoted by an arbitrary symbol η. Thus, each set of l-value has a set of zeros, and each of them corresponds to a value for k given by (4.4). From figure 4.1, one obtains for $j_0(kr), l = 0$ values of η^2 for which Bessel function value is zero. The orbital angular momentum $l = 0$ is labeled as s-wave, $l = 1$ as p-wave, and $l = 2$ as d-wave. One gets $\eta^2 = 9.87$ (1s), 39.48 (2s), 88.83 (3s), $\eta^2 = 20.14$ (1p), 59.68 (2p), 118.9 (3p), and $\eta^2 = 33.21$ (1d), 82.72 (2d), etc.

$$kr_0 = \eta \quad or \quad k^2 = \frac{\eta^2}{r_0^2}$$

Thus, energy of a given level is obtained by substituting the value of η^2 in units of $\hbar^2 / 2mr_0^2$, and it is given as

$$E = \frac{\hbar^2 k^2}{2m} = \left(\frac{\hbar^2}{2mr_0^2} \right) \eta^2 \qquad (4.5)$$

Energy eigenvalues are both functions of total quantum number n and the angular momentum quantum number l. Regardless of the form of potential, the quantum numbers representing nucleons are n, l, j, m.

This problem is made mathematically simpler if one assumes a potential well with infinite depth and potential form given as

$$V(r) = \infty \quad r > r_0$$

The solution of the radial wave equation for the infinite square well gives the energies of levels from the bottom of the well in the following order:

1s 1p 1d 2s 1f 2p 1g 2d 1h 3s 2f 1i 3p

where letters s, p, d, etc., represent $l = 0, 1, 2$, etc.

Total number of protons or neutron in a given (l) state with spin $\pm\dfrac{1}{2}$

and m_l as -1, . . . ,+l is $2(2l+1)$; this will predict the shell closures at 2, 8, 18, 20, 34, 40, 58, etc., for $l = 0, l = 1, l = 2, l = 3,$ and $l = 4$, which are not consistent with known magic numbers of nuclei.

4.8. Shell Theory Potential

Some authors have used a modified average potential roughly proportional to the density of the nucleons in the nucleus and has the form given as

$$V = -V_0 \frac{1}{1+\exp\left(\dfrac{r-R}{a}\right)} \tag{4.6}$$

where a is the surface thickness taken as equal to 0.65 fm, R is the nuclear radius. Values of V_0 and R are determined from elastic scattering cross section as

$$V_0 = -57 Mev$$

$$R = r_0 A^{\frac{1}{3}} \ fm$$

A correction to the value of V_0 is due to symmetry energy, which arises from unequal number of neutrons and protons in a nucleus.

The symmetry potential is approximately as

$$\Delta V_s = \pm 27\, Mev \times \frac{N-Z}{A} \begin{pmatrix} - \ neutrons \\ + \ protons \end{pmatrix} \tag{4.7}$$

Shell theory potential for a proton must also include a term due to Coulomb repulsion given as

$$V(r) = \frac{Ze^2}{4\pi\varepsilon_0 R_c}\left\{1+\frac{1}{2}(1-\frac{r}{R_c})^2\right\}, r < R_c$$

$$V(r)_{p-p} = \frac{Ze^2}{4\pi\varepsilon_0 r} \quad for\ r > R_c$$

(4.8)

where R_c is the nuclear radius.

For a complete description of the average potential, one must include the following terms in the potential.

1. Spin dependence (exchange force).
2. Parity dependence (exchange force).
3. Noncentral or tensor force. One can ignore this for average potential.
4. Velocity dependent forces (V_0).

V_0 is different for protons and for neutrons. Analysis of V_0 for protons in the energy range of $9 < E_p < 22 Mev$ gives

$$V_0(Mev) = \left[53.3 - 0.55E + 0.4\frac{Z}{A^{\frac{1}{3}}} + 27\frac{N-Z}{A}\right]$$

(4.9)

The second term in the equation is energy dependent term. Similar potential shapes with slightly different values are used by different authors to calculate the properties of nuclei such as binding energies, magnetic moments, and energies of excited states of nuclei.

4.9. Spin-orbit Interaction

In 1949, Mayer et al. proposed that spin-orbit interaction will produce correct magic numbers. The interaction between the spin (s) of a particle with its angular momentum (l) gives rise to a potential given as

$$-V_{l,s}(r)\,\vec{l}.\vec{s}$$

$V_{ls}(r) = -f(r)(l.s)$, $f(r)$ has Wood-Saxon form.

Such a potential can cause a splitting of energy levels depending upon the relative orientation of l and s.

l and s can couple to give the total angular momentum j as

$$\vec{j} = \vec{l} \pm \vec{s}$$

From the cosine law, one gets

$$\vec{l}.\vec{s} = \frac{1}{2}\left(j^2 - l^2 - s^2\right) = \begin{cases} \dfrac{1}{2}l & for\ j = l + \dfrac{1}{2} \\[2mm] -\dfrac{1}{2}(l+1) & for\ j = l - \dfrac{1}{2} \end{cases} \qquad (10)$$

$$j^2 = J(j+1)$$
$$l^2 = l(l+1)$$
$$s^2 = s(s+1)$$

Thus, this spin-orbit potential depends upon l-value as well as it is positive when l and s are parallel and negative when l and s are antiparallel, i.e., potential energy decreases when l and s are parallel and increases when l and s are antiparallel. The difference in energy of levels due to this interaction is

$$\Delta V_{S.L} = \frac{1}{2}f(r)(2l+1),$$

where $f(r)$ is determined by comparing it with the experimental data.

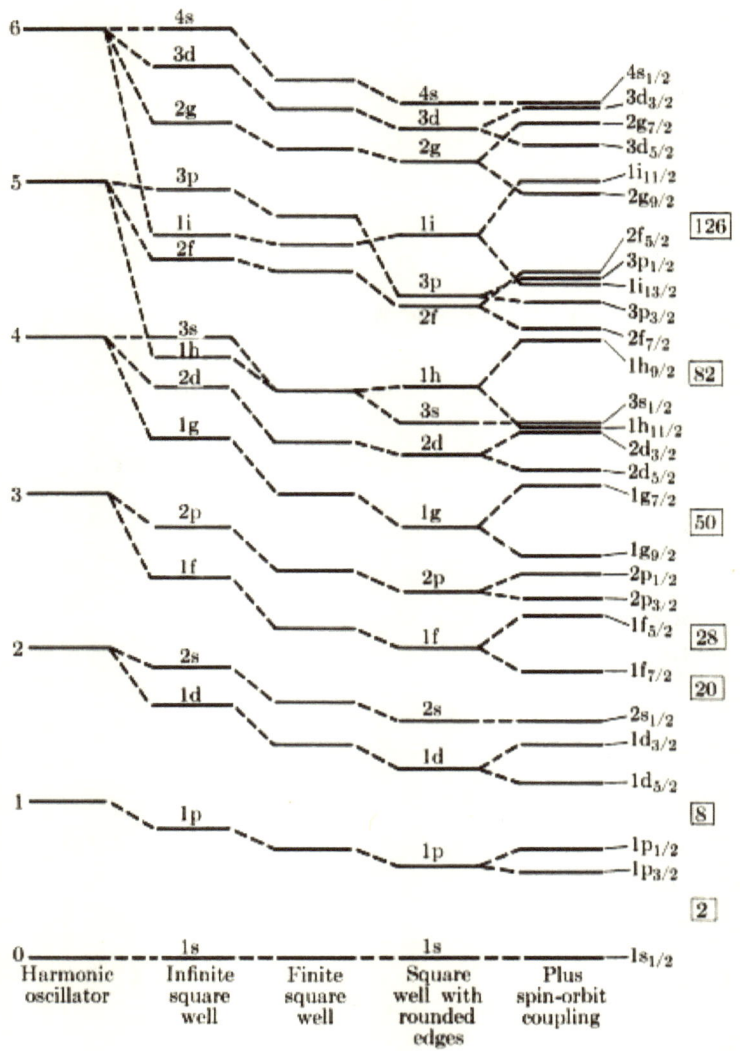

Figure 4.2. Order of energy levels according to the independent-particle model with various assumptions for the shape of the nuclear potential. [From B. T. Feld, *Ann. Rev. Nucl. Sci.* **2,** 249 (1953). Reproduced by permission.]

As a result of the two possible orientation of the nucleon spin, a level can accommodate $2(2j+1)$ protons or neutrons. In view of the charge independence of nuclear force, the energy levels for neutrons are similar to those of protons except for the effect of Coulomb force between protons.

The inclusion of $l.s$ potential by Mayer, Haxel, Jensen, and Suess modifies the level energies as are shown in figure 4.2 (column 5). This produced the correct magic numbers with shell closures at proton and neutron numbers of 2, 8, 20, 28, 50, 82, and 126. These numbers are consistent with the experimental data of relative abundance of nuclei.

4.10. Angular Momentum of Nuclear Ground States

In shell model terminology, each level is characterized by a principal quantum number n, an l-value, and a j-value ($l \pm \frac{1}{2}$). We represent such quantum numbers as configuration.

4.11. Definitions of Configuration, Term, and State

Configuration is a listing of what orbits are occupied by particles without regard to m_j values. For example, $\left(d_{\frac{5}{2}}\right)^2$ is a configuration. It represents 2 neutrons in a $d_{5/2}$ orbit for O_8^{18} nucleus. The sets $(d_{5/2}, s_{1/2}), (d_{5/2}, d_{3/2})$ are also configurations, which represent one neutron in $d_{5/2}$ and one neutron in $s_{1/2}$ orbit or one neutron in $d_{5/2}$ and one neutron in $d_{3/2}$ orbits.

The energy of a state with any configuration is the sum of the energies of all occupied orbits. For example, the energy for $\left(d_{\frac{5}{2}}, d_{\frac{3}{2}}\right)$ configuration will be the sum of energies of one particle in $d_{5/2}$ orbit and one particle in $d_{3/2}$ orbit.

A *term* is a configuration with total angular momentum coupling specified. For example,

$$\left(d_{\frac{5}{2}}\right)^2_{I=0} \text{ or } \left(d_{\frac{5}{2}}\right)^2_{I=2} \text{ or } \left(d_{\frac{5}{2}}, d_{\frac{3}{2}}\right)^2_{I=2} \text{ is a term.}$$

These terms specify that two particles in $d_{5/2}$ orbit can couple to produce total angular momentum of 0, 2, and 2 respectively.

A *state* is the solution of the Schrödinger equation for the system. It has a definite energy, a definite wave function, a definite total angular momentum, a definite parity, and values of other definite properties such as magnetic and quadrupole moments (E, ψ, J, π, μ, Q, etc.).

4.12. General Observations of Nuclear Properties

1. All even-even nuclei with even Z and even N have $J^\pi = O^+$. There are no known exceptions to this.
2. An odd nucleus with odd Z and even N or odd N with even Z will have a total J^π equal to the half integral angular momentum J and parity $(-1)^l$ of the unpaired odd particle. There are few exceptions to this rule.
3. An odd-odd nucleus with Z odd and N odd will have a total angular momentum, which is the vector sum of total angular momentum of odd neutron j_n and odd proton j_p such that

$$\left| j_n + j_p \right| \le J \le \left| j_n - j_p \right|$$

and parity $\pi = (-1)^{l_n + l_p}$

4.13. Intrinsic Spin and Magnetic Moment (μ) of a Nucleus and the Shell Model

Each nucleus has an intrinsic spin known as J, which is the vector sum of angular moment L and spin S for all the nucleons of the nucleus. For even mass nucleus, it is an integral multiple of $\hbar$ and for odd mass nucleus it is a multiple of $\dfrac{\hbar}{2}$.

Since protons inside the nucleus have a charge e and are moving with a velocity in orbits, they give rise to magnetic moments whose values are given as

$$\mu_l = \frac{1}{2}\frac{e\hbar}{M_p c}\int \psi^* \left(\sum_{k=1}^{Z} l_k \right)\psi d\tau \qquad (4.11)$$

where l_k is the angular momentum of the kth proton. The sum is over all protons (Z).

One can thus write

$$\mu_l = \mu_0 \sum_{k=1}^{Z} g_k l_k$$

$$\mu_0 = \left(\frac{e\hbar}{2M_p c}\right) \text{ in nuclear magneton}$$

Each proton also has an intrinsic spin s in addition to l and give rise to a magnetic moment due to its spin. This is given as

$$\mu_s = \mu_0 \sum_{k=1}^{z} g_k s_k$$

A magnetic moment operator of a nucleus is due to the combined effects of orbital and spin moments. This is written as M for a nucleus with N nucleons.

$$M = \mu_0 \sum_{k=1}^{N} \left(g_k^l l_k + g_k^s s_k \right) \qquad (4.12)$$

where g_k^l is the orbital factor for a nucleon with angular momentum l.

$g_k^l = 1$ for protons, $g_k^l = 0$ for neutrons
g^s is the spin factor for nucleon with spin s. Its value for
$g_k^s = 5.5856$ for protons
$g_k^s = -3.8263$ for neutrons

The magnetic moment is due to motion of charge only.

The magnetic moments of various nucleons and electrons have been measured precisely. The experimental value are given as

μ_e = -1.011 Bohr magneton

μ_p = 2.79275 nm

μ_n = -1.9135 nm

In odd nucleus, according to the shell model, all except one nucleon will form pairs with antiparallel j-vector. Hence, the total angular momentum J of the nucleus is equal to the angular momentum j of the last unpaired nucleon.

Hence, the magnetic dipole moment of a nucleus, according to shell model, is produced by odd nucleon.

If the last nucleon is in an $l = 0$ or s-state, magnetic dipole moment will be equal to that associated with the odd nucleon. If the odd nucleon is a proton, then μ = 2.79275 nm, and if the odd nucleon is a neutron μ = -1.913nm.

Magnetic moment is observable when the nucleus is placed in an external magnetic field.

The sum of the components g_l and g_s vector gives vector g_j, whose component along Z direction gives the observable interaction of magnetic moment with the applied magnetic field.

From the above figure, one sees the g_j is given as

$$g_j \sqrt{j(j+1)} = g_l \sqrt{l(l+1)} \cos(l, j) + g_s \sqrt{s(s+1)} \cos(s, j) \qquad (4.13)$$

From the cosine rule of the triangle formed by l, s, and j, one gets

$$\cos(l, j) = \frac{j(j+1) + l(l+1) - s(s+1)}{2\sqrt{l(l+1) j(j+1)}}$$

$$\cos(s, j) = \frac{j(j+1) + s(s+1) - l(l+1)}{2\sqrt{s(s+1) j(j+1)}}$$

Inserting these values in the above equation 4.13, one gets

$$g_j = \frac{[j(j+1)+l(l+1)-s(s+1)]g_l}{2j(j+1)} + \frac{[j(j)+1)+s(s+1)-l(l+3)]g_s}{2j(j+1)}$$

(4.14)

For a single particle, the spin is $s = \pm\dfrac{1}{2}$, j can have two values $j = l + \dfrac{1}{2}$ and $j = l - \dfrac{1}{2}$.

One has given names for the case when spin vector $s = \frac{1}{2}$ is antiparallel or is parallel to l vector. These names are as follows:

$J = j = l + \dfrac{1}{2}$, stretch case

$J = j = l - \dfrac{1}{2}$, jackknife case

j can be also expressed as J since the j value is for a single nucleon. Thus, the magnetic dipole moment along J-vector is given as

$$\mu = g\sqrt{J(J+1)}\left(\frac{J}{\sqrt{J(J+1)}}\right) = gJ$$

(4.15)

One then gets from equation 4.13. Values of magnetic moment for both stretch and jackknife cases are as follows:

$$\mu = \frac{1}{2}g_s + \left(J - \frac{1}{2}g_l\right)$$

(4.16)

$$\mu = \frac{1}{J+1}\left[\frac{1}{2}g_s + \left(J + \frac{3}{2}\right)g_l\right]$$

(4.17)

Values of μ for $J = \dfrac{1}{2}, \dfrac{3}{2}, \dfrac{5}{2}, \dfrac{7}{2}$, and $9/2$ for odd proton Z and for odd neutron have been calculated based upon the equations 4.16 and 4.17 and compared with experimental results.

Experimental measurements of magnetic dipole moments of nuclei with odd proton is shown in figure 4.3 and that of odd neutron is shown in figure 4.4. Solid lines are Schmidt lines for stretch and jackknife cases.

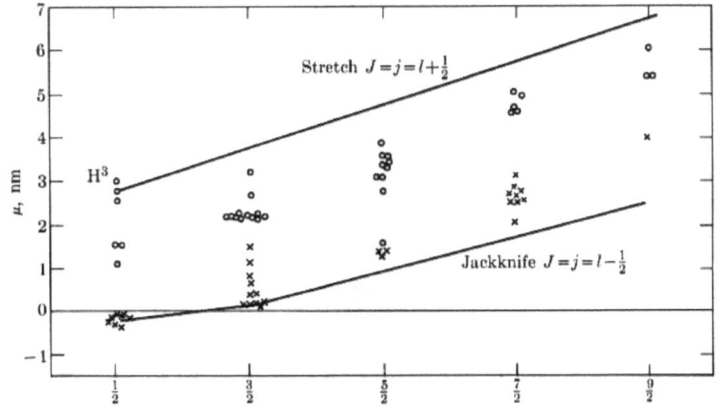

Figure 4.3 Experimental magnetic dipole moments and Schmidt lines for nuclei of odd Z.

A comparison of measured and calculated values of dipole moments based on the shell model of some nuclei are given in table 4.1. Discrepancy between the two is small.

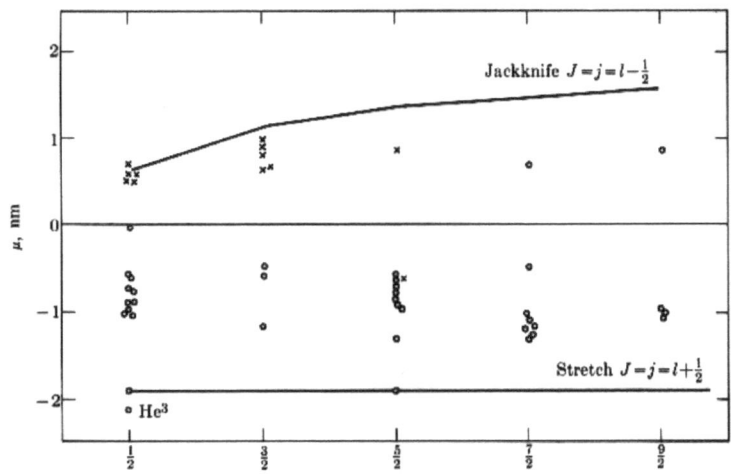

Figure 4.4 Experimental magnetic dipole moments and Schmidt lines for nuclei of odd N.

Table 4.1. Observed values and shell-model predicted values of magnetic moment of some nuclei

Nucleus	spin	μ(observed)	μ(Theory)
n	$\frac{1}{2}$		-1.91
H^1	$\frac{1}{2}$		2.79
H^2	1	0.86	0.88
H^3	$\frac{1}{2}$	2.98	2.79
He^3	$\frac{1}{2}$	2.13	-1.91
He^4	Closed Shell		
Li^6	1	0.82	0.88
Li^7	$\frac{3}{2}$	3.26	3.07
Be^9	$\frac{3}{2}$	-1.17	-1.14
C^{12}	Closed Shell		
Mg^{25}	$\frac{5}{2}$	- 0.85	-1.0

4.14. Electric Quadrupole Moments of Nuclei

Electric quadrupole moments of nuclei arise due to the deviation of charge distribution of the nucleus from a spherical charge distribution. Electric quadrupole moment of a nucleus is given as

$$Q = 1/e \ (3z^2 - r^2)\rho \ dv$$

If a nucleus is viewed as a uniformly charged ellipsoid of revolution with the symmetry axis along the total angular momentum J vector, the quadrupole moment will be given as

$$Q_J = 6/5\ Z\ R^2\ (\Delta R/R)$$

where Z is the atomic number and R is the average radius. Hence, values of Q are expressed in barns in units of 10^{-24} cm^2.

According to the above equation, $Q = 0$ is for a spherically symmetric charge distribution.

A charge distribution stretched along the z direction (prolate shape) will produce a positive value of Q, and a charge distribution stretched along the x-y plane (oblate shape) will produce a negative value of Q.

Nuclei with a single nucleon such as a neutron or a proton outside the spherical core can give rise to a nonspherical charge distribution and thus can acquire electric quadrupole moments.

Electric quadrupole moments of nuclei have been measured extensively. These moments of ground states of nuclei exhibit a definite trend when plotted against Z or N.

Figure 4.5 shows values of electric quadrupole moments of odd A nuclei against Z or N, whichever is odd.

Quadrupole moment falls sharply through zero for certain values of nucleon numbers representing doubly closed shells.

For certain nuclei in the rare earth regions such as Eu, Cd, Ho, Lu, Ta, Ir, etc., the quadrupole moments are very large as seen in the figure. Quadrupole moment for Er^{167} has the largest value of about 8 barns.

Similarly, some medium weight nuclei such as Mn^{55} and Co^{69}, and some light nuclei such as B^{10}, B^{11}, and Be^9 have large quadrupole moments. These nuclei are most likely deformed from spherical shapes.

In 1950, these values of quadrupole moments were reported by scientist, Townes, in a seminar at Columbia University. James Rainwater offered an explanation for these large values of quadrupole moments.

According to Rainwater (4.8), a large number of protons outside the closed shell in such nuclei can polarize the nuclei to be permanently deformed from a spherical shape. This could produce an asymmetrical charge distribution to give rise to the electric quadrupole moment. These ideas were further developed by A. Bohr and B. Mottelson (4.9) to investigate nuclear properties of deformed nuclei. Rainwater, Bohr, and Mottelson received Nobel Prizes for the idea and for the development of the collective model.

Shell model is unable to predict the correct values of electric moments.

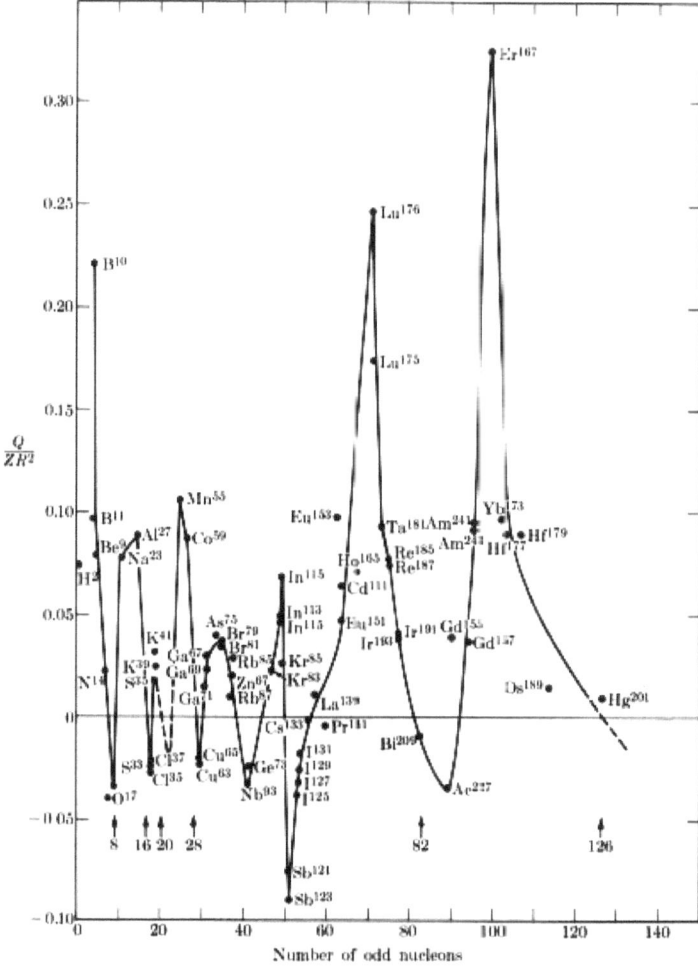

Figure 4.5   Quadrupole distortion $Q/ZR^2 (\approx \Delta R/R)$ for odd-A nuclei plotted vs. the odd nucleon number Z or N. (From E. Segrè. *Nuclei and Particles*, New York: W. A. Benjamin, Inc., 1964 Reproduced by permission.)

4.15. Excited States of Nuclei and Shell Model

In atomic theory, it is easy to calculate the excited states of atoms when they absorb energy by calculating the total energy of electrons in different allowed orbits. However, the excited states of nuclei when they absorb energy after interaction with other nucleons or gamma rays have a much more complex structure. The reason being that the nucleon-nucleon interaction is very complicated, involving excitations of several nucleons in one or many orbits.

For light nuclei, it is assumed that excited states are formed due to the motion of a single nucleon to higher orbits. Hence, one can predict energies of excited states by calculating the energies of nucleons in different orbits with different values of angular momenta and parities using shell-model potential.

Simplest nucleus deuterium or tritium have no excited states. The He_2^4 nucleus has an excited state at about 16.9 MeV. As the nucleus have more nucleons, they can be excited to higher states, which can be reached by nuclear reactions. The energies and J^π values, etc., can be determined and then compared with the shell-model calculations.

Level energies of nuclei O_8^{16} and O_8^{18} taken from Irvine (4.10) are given in figure 4.6.

These are even even and closed shell nuclei. Their spectra are simple. In O_8^{16}, the first excited state observed at excitation energy of 6.06 MeV has $J^\pi = 3^-$. Other excited states are seen up to about 26 MeV. Similarly, in O_8^{18}, the first excited state is at 1.982 MeV, and many other states are observed up to about 8.8 MeV. Spins and parities J^π of many of these states are also known as were determined from energies of γ- rays.

In comparison, an odd even nucleus F_9^{19}, which has 9 protons and 10 neutrons show much more complex spectra of excited states. Number of excited states keep on increasing as we go to heavier masses except in certain mass region near closed shell nuclei. The $_{20}Ca^{40}$ spectra is very simple with fewer excited states.

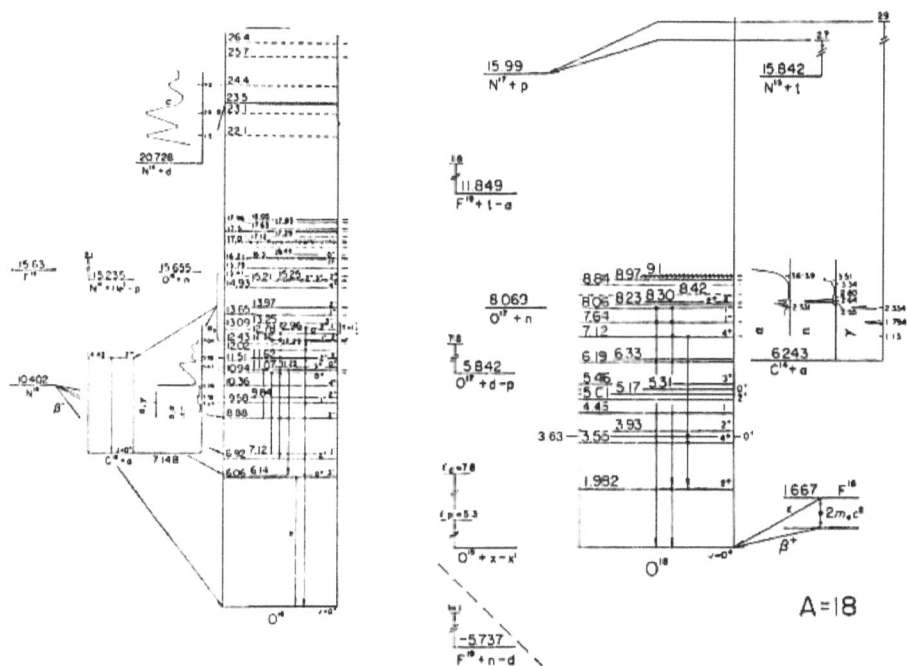

Figure 4.6. Level Energies for Nuclei O ^{16} and O ^{18}

The majority of the excited states based upon shell model are due to a single or few nucleons moving in an average spherically symmetric potential as described earlier.

According to the shell model, the distribution of nucleons is governed by quantum numbers. For example, the first shell $N = 1$ has 2 protons and 2 neutrons. The second shell $(P_{\frac{3}{2}}, P_{\frac{1}{2}})$ will have 6 protons and 6 neutrons. Third shell $N = 3$ will have $(d_{\frac{5}{2}}, s_{\frac{1}{2}}, d_{\frac{3}{2}})$ 12 protons and 12 neutrons. The energies of nucleons in $N = 3$ shell being in $d_{\frac{5}{2}}, s_{\frac{1}{2}}$ and $d_{\frac{3}{2}}$ states are shown in the figure 4.3.

Energy of a state is much more complicated as they take into account the effect of short range residual interactions (v_{ij}), and the corresponding energies show deviation from the average energy of the configuration. In most cases, such deviations are small.

Take the case of O_8^{18}. This nucleus has 8 protons and 10 neutrons. The last two neutrons are in $d_{\frac{5}{2}}$ ground state with $J^\pi = O^+$. When the nucleus is excited, the energy is absorbed by a single neutron, which can move to the next higher state of $s_{1/2}$, leaving the other neutron in $d_{\frac{5}{2}}$ state.

Simple single nucleon theory cannot predict properties of highly excited states. For the excited states above few MeV in nuclei, one has to assume that these $J^\pi = \frac{5}{2} \pm \frac{1}{2}$ or $J = 2, 3, 4$, and positive parity. The energy of the excited state would be the energy of $s_{\frac{1}{2}}$ state.

States are produced by many nucleons. With the development of very high speed large computers, it has become possible to do large-scale shell model diagonalization of interaction matrices to predict energies of such states. At the moment, one is not able to predict the properties of highly excited states in heavy nuclei where the spacing between levels is only few electron volts, the level density is very high.

The spin and parity will be determined by combining spins of two nucleons and parity is $(-1)^{l_1+l_2}$. Shell theory is an approximation. It is impossible to account for the complex interactions with a simple potential well. When nucleons collide, the interaction gives rise to residual interaction, which are small and are taken as perturbations changing the orbit with the application of conservation laws of shell numbers, angular momentum, parity, etc.

Over the years, many scientists have calculated the excited level energies of light nuclei up to several MeV and compared with the experimental data. By and large the agreement has been satisfactory for light nuclei.

4.16. Summary of Shell Model

Hence, the simple shell model picture including spin-orbit coupling predicts good agreement with properties of nuclei with known data such as in the following:

1. Correct magic numbers (2, 8, 20, 28, 50, 82, 126)
2. Spins and parities of low lying states in light mass nuclei
3. Binding energy of nuclei
4. Cross sections of neutron capture by nuclei

5. Magnetic dipole moments of nuclei with small deviations from the experimental values
6. Transition probabilities of emission of gamma rays emitted from nuclei

But it fails to give correct values of quadrupole moments of nuclei. Shell model predicts these moments to be zero based upon spherical nuclei. This model in its simple form treats the individual nucleons to move in stationary orbits and are paired off in such a way that the values of many nuclear parameters are determined solely by a single unpaired nucleus. This model ignores any correlated or collective motion of several nucleons—a realistic picture. Hence, the shell model is of limited validity, mostly useful for light nuclei $< Ne^{2c}$.

4.17. Models of Deformed Nuclei

In the shell model, the nuclear potential is treated as spherically symmetric and stationary.

We discuss here the case where the nucleus is no longer spherical but is deformed. This situation arises when the shells in some nuclei are not completely full. Such nuclei show large values of electric quadrupole moments as shown in figure 4.4.

Historically, it is known that the concept of collective motion of nucleons in nuclei was first proposed by James Rainwater (4.8) to explain the observed large electric quadrupole moments in nuclei in the atomic mass region of 150-190 known as rare earth nuclei.

In such cases, nuclear shapes are deformed into ellipsoids, or the nucleus is known as a spheroid, i.e., two of the three principal axes are equal. In cases where the third unequal axis is larger than the others, the nucleus takes the shape of a football called prolate shape. In cases where the unequal axis is shorter than the two equal axis, the nucleus has a shape of pumpkin and is known as oblate or a cigar shape.

Deformation of nucleus is defined by a deformation parameter δ given as

$$\delta = \frac{R_3 - R_1}{\overline{R}}$$

(4.18)

where R_3 is the unequal axis and $R_1 = R_2$. The δ-values can vary from zero to values of $\delta = 0.5$ or 0.6 for large deformation.

4.18. Rotational Model of the Nuclei

When the nuclei are deformed, they are set into rotational motion with angular velocity w or the nuclei vibrate about their axes of symmetry.

Nuclear rotational motion is somewhat complex than the rotation of a rigid body. Both vibration and rotational motions involve orderly displacements of many nucleons and are known as collective motion of nuclei where all particles are participating in collective motion.

The rare earth nuclei are found to have large quadrupole moments due to large deformations of the nucleus in the form of a shape of a football.

In this, the nonspherical surface of the nucleus will rotate about all axis of symmetry. For even-even nucleus, the entire nucleus rotates about an axis known as the symmetry axis.

From mechanics, we know that a rotating body with an angular velocity w and a moment of inertia I has rotational kinetic energy given as

$$E_{rot} = \frac{1}{2} I w^2 \tag{4.19}$$

The rotating body also has an angular momentum L, and it is given as

$$L = I w \tag{4.20}$$

In the case of nuclear motion, J is the total angular momentum about the axis of symmetry. Hence,

$$J = I w$$

Substituting the value of $w = \dfrac{J}{I}$ in the energy equation 4.19, we get

$$E_{rot} = \frac{1}{2} I \left(\frac{J}{I} \right)^2$$

$$E_{rot} = \frac{1}{2} \frac{J^2}{I} \tag{4.21}$$

Total angular momentum J has values as

$$J = \sqrt{J(J+1)} \hbar$$

Substituting this into the equation 4.19 above, one has

$$E_{rot} = \frac{J(J+1)\hbar^2}{2I} \tag{4.22}$$

Therefore, if the nucleus behaves as a rigid body with moment of inertia I, its energy of excitation depends upon J-value in units of $\dfrac{\hbar^2}{2I}$.

To a first approximation, if the nucleus is assumed to be spherical in shape, the moment of inertia of a solid sphere rotating about an axis passing through its center is given as

$$I = \frac{2}{5} MR^2$$

where M is the mass of sphere and R its radius.

Applying this to a spherical nucleus, mass of nucleus is $= (Am_n)$ where A is the number of nucleons and m_n is the mass of one nucleon. Radius of the nucleus is given as $R = r_o A^3$ where r_o is a constant and A is the atomic mass. Thus, moment of inertia for a rigid nucleus is

$$I = \frac{2}{5}(Am_n)r_0^2 A^{\frac{2}{3}} \tag{4.23}$$

Energies are expressed in units of $\dfrac{\hbar^2}{2I}$

$$E^* = \frac{5\hbar^2}{4(Am_n)r_0^2 A^{\frac{2}{3}}} \tag{4.24}$$

Substituting values of A, $r_0, \hbar, m_n$ in equation 4.24, one gets

$$E^* = 21.2 \frac{J(J+1)}{A^{\frac{5}{3}}} \text{ MeV} \tag{4.25}$$

Hence, the excited state energies of deformed even-even rotating nucleus will be given as follows: the ground state, the lowest energy state has $J = 0$. The J values of excited states are 2, 4, 6, 8, etc. The energies of such excited states due to equation 4.22 are given below:

$$E_2 = 6\frac{\hbar^2}{2I} \text{ for } J = 2 \tag{4.26}$$

$$E_4 = 20\frac{\hbar^2}{2I} \text{ for } J = 4 \tag{4.27}$$

$$E_6 = 42\frac{\hbar^2}{2I} \text{ for } J = 6 \text{ and so on} \tag{4.28}$$

The figure 4.7 below due to Bohr (4.11) shows the observed energies, angular momentum, and parities of a deformed even-even nucleus Hf^{180}.

	Experimental	Theoretical I	Theoretical II
8^+	1085.3	1119.6	1085.4
6^+	641.7	653.1	642.0
4^+	309.3	311.0	308.9
2^+	93.3	(93.3)	93.2
0^+	0	0	0
$_{72}Hf^{180}$			

Figure (4.7) Energy-level diagram of Hf^{180} with experimental and calculated excitation energies. [From A. Bohr, "Rotational States of Atomic Nuclei," *Mat. Fys. Skr. Dan. Vid. Selsk.* **1**, No. 8 (1959).]

The agreement of energies between the experiment (column 1) and calculated values (column 2) based upon the equations 4.26, 4.27, and 4.28 is excellent. A small discrepancy in energies can be explained as resulting from an increase in the moment of inertia with increasing value of *J*.

When a correction term due to the above is added, one obtains for energies of the excited states given as

$$E_J = \frac{\hbar^2}{2I} J(J+1) - BJ^2 (J+1)^2 \tag{4.29}$$

Numerical value of *B* is adjusted based upon the discrepancy of the experimental data with the theory.

This correction makes the agreement excellent with most data as is shown in the column 3 of figure 4.7 labeled as theoretical II.

Measurements of excited state energies and *J* values for even- even deformed spheroidal nuclei Yb^{166}, Hf^{170}, W^{172}, and Os^{182} are shown in figure 4.8. In each of these cases, one observes a regular pattern of excited states with $J^{\pi} = 2^+, 4^+, 6^+, 8^+$, etc., and their energies are as predicted by the rotational model given by equation 4.29.

High angular momentum states with $J = 10, 12, 14$, and 16 are also observed, and their energies agree with the prediction of the rotational model.

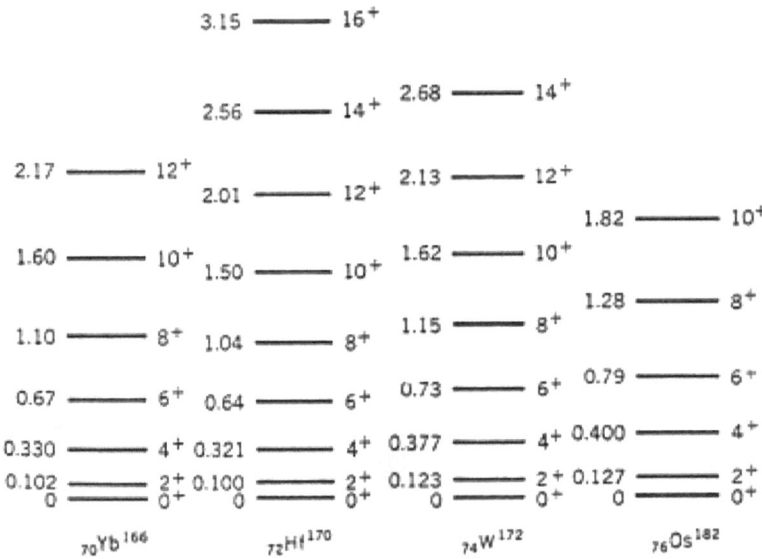

Figure (4.8) Experimentally determined ground-state rotational bands in typical even-even spheroidal nuclei. The numbers at the left are excitation energies in MeV (top row) or keV (bottom row), and I^{π} for the states is shown at the right. These states were determined in coulomb excitation experiments

Ratio of rotational state energies of $J = 4, 6, 8$, and 10 to the energy of 1st excited state $J = 2$ compiled by Nathan et al. (4.12) are plotted in figure 4.9 for a number of even-even nuclei in the mass region of 150-250 and are compared with the prediction of the formula (4.29) shown as straight lines. The agreement between the two is excellent for the range of deformed nuclei particularly nuclei in the rare earth region with A values from 150 to 180. Small discrepancy between experimental values and theoretical calculation are explained earlier.

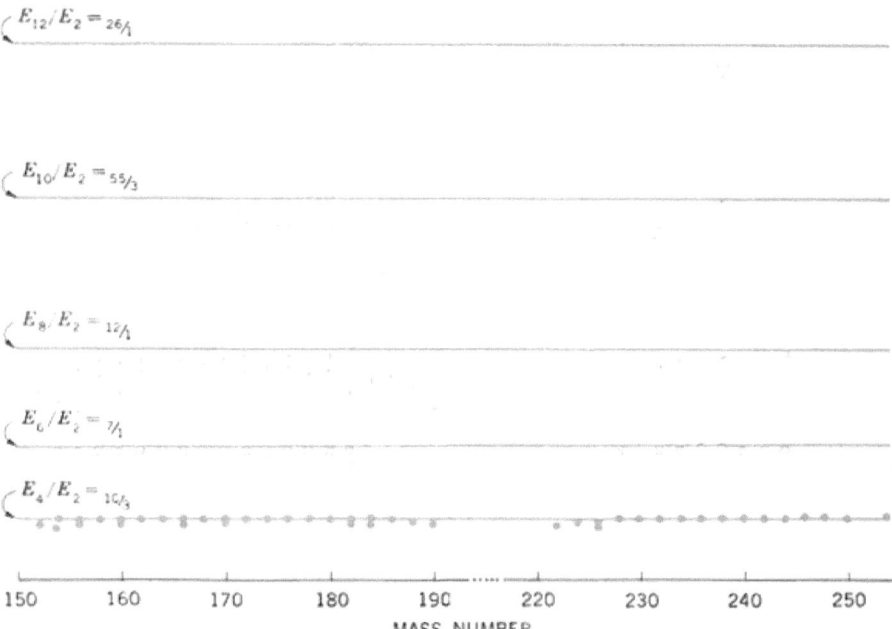

Figure 4.9 Ratio of energies of various members of ground-state rotational bands in spheroidal even-even nuclei to the energy of the lowest (2⁺) member. Subscripts are l of the states, and the horizontal lines are the predictions of (6-7). [*From O. Nathan and S. G. Nilsson in K. Siegbahn (ed.), "Alpha, Beta, Gamma Ray Spectroscopy," North-Holland Publishing Company, Amsterdam, 1965; by permission.*]

4.19. Odd Nuclei in a Deformed Well

The simplest states in odd—A nonspherical nuclei are the single quasi particle state. When a nucleus is regarded as a rotating ellipsoid with an axis of symmetry Z', additional quantum numbers are needed for a complete specification of its state.

The total angular momentum of individual particle j is no longer conserved, but its projection onto the symmetry axis is conserved. Let Z' be the symmetry axis of the deformed nucleus as shown in figure 4.10.

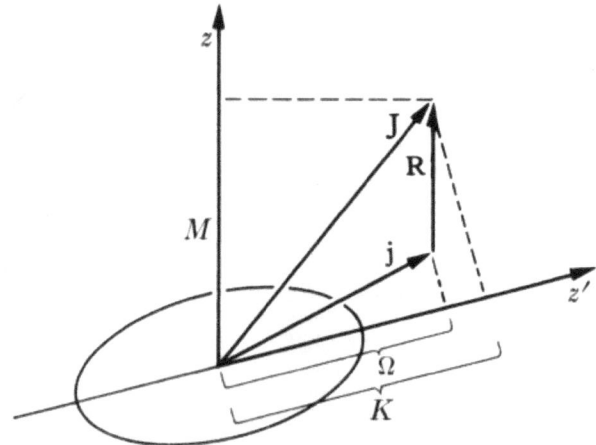

Figure 4.10   Coupling scheme for angular momenta of a single particle interacting with a spheroidal shell of nuclear surface.

Coupling of the total angular momentum of an odd individual particle j with the total angular momentum R of the even-even core known as rotator as a whole gives the angular momentum J of the odd-odd nucleus as shown above.

Ω is the projection of j along Z' axis,

K is the projection of J along Z' axis

where $j + R = J$ (as shown in the figure above).

Total Hamiltonian for such a deformed nucleus is given as

$$H = H_{rot} + H_p + V(R)$$

H_{rot} = Hamiltonian for the rotator

H_p = Hamiltonian for the odd particle

$V(r)$ = Interaction potential between the particle and rotator

Just as in the case of spherical nuclei, the simplest states in odd—A spheroidal nuclei are the single-quasi particle states, those which differ from the ground states of neighboring even-even nuclei by the addition of one particle or one hole.

For even-even nuclei, the total angular momentum is zero; the angular momentum of the odd nucleus comes from the configuration of the odd particle or hole.

If the j of the odd particle is known, the ground state total angular momentum is $K = j + R$ and total spins of excited states are $J = K + 1$, $K + 2$, $K + 3$, and energies of these states are given as

$$E_J = \frac{\hbar^2}{2I}\left[J(J+1) - K^2\right] \tag{4.30}$$

where I is the moment of inertia of the deformed odd nucleus and

K is the projection of j on the axis of symmetry Z. J must be greater than or equal to K. Lowest energy state of a deformed nucleus is $J = K$.

Since the odd nucleus can be in any orbit producing different values of j, its projection along the symmetry axis K can take different values. Hence, there can be many excited states built on different values of K. A level energy diagram of even odd nucleus Yb[169] due to Harmatz et al. (4.13) and Wilson et al. (4.14) is shown in the figure 4.11. This shows rotational states corresponding to three K bands. Levels built on ground state band $K = 7/2^+$ with excited states of $J = 9/2$, $11/2$, and $13/2$ and positive parities and levels built on excited state band $K = 1/2$—with excited states $3/2$, $5/2$, $7/2$, $9/2$, and $11/2$ and negative parities and second excited state band $K = 5/2$ and corresponding excited levels at $7/2$, $9/2$, and $11/2$ and negative parities. Level energies are given by the rotational model formula (4.30) for odd-even nucleus.

A complete discussion of single particle state in a deformed nucleus is given later by the unified model.

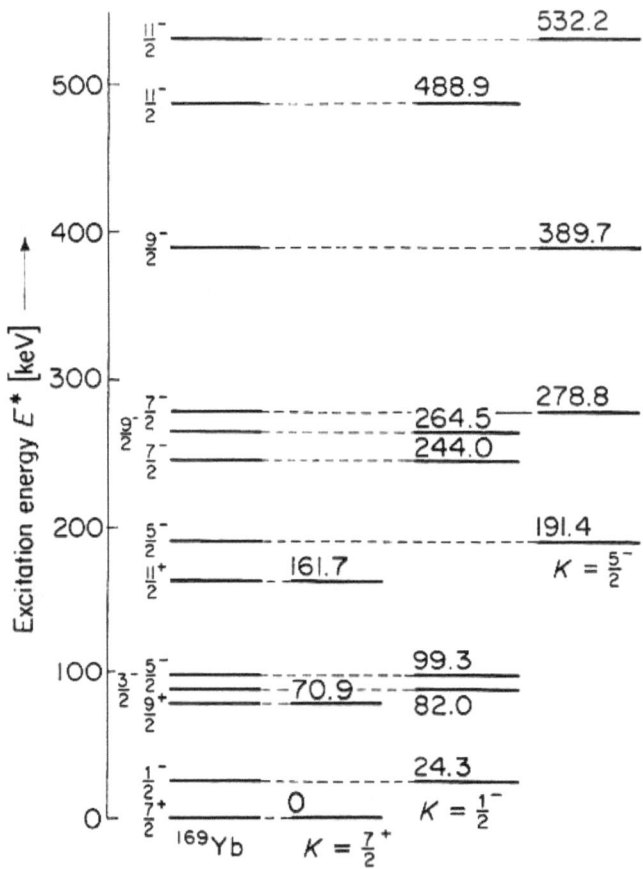

Figure 4.11. Energy level spectrum of Yb[169] into three rotational bands characterized by the quantum numbers $K = 7/2^{+}$, $1/2^{-}$, and $5/2^{-}$, respectively. This level scheme is obtained from the study of electron capture decay of Lu[169].

4.20. Vibration States of Even-even Nuclei

Vibration states have been observed in molecular spectra. Energies associated with the collective motion of an incompressible liquid were deduced by Raleigh. Bohr developed the nuclear hydrodynamic model in a general manner. When a nucleus is not very much deformed or is even closed to spherical shape, the nuclear surface can undergo surface vibrations or oscillations similar to a molecule except the energies of the nuclear surface vibrations are much greater than those of a molecule.

Nuclear surface vibrations produce collective motion of the nucleus as a whole. This motion consists of a surface wave going round the nucleus carrying a certain amount of mass that produces a pressure on the bulge because of the centrifugal force.

This pressure is counteracted by the surface tension representing a restoring force.

Quantum number representing the surface vibration is known as phonon, which carries an angular momentum $\lambda = 2, 3, 4, \ldots$, etc. and the parity $\pi = (-1)^{\lambda}$.

The $\lambda = 1$ vibration known as dipole vibration gives a net displacement of the center of mass and thus does not produce surface vibration. The next lowest made $\lambda = 2$ known as quadrupole vibration produces the surface vibration as shown below.

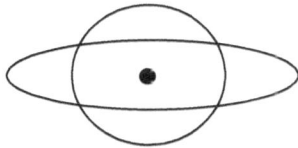

The mode with $\lambda = 3$ produces octupole surface vibration as shown below.

The energy related to surface vibration is expressed in units of angular frequency ω as for a harmonic oscillator. The angular momentum of state is expressed in terms of λ value known as phonon. Thus, a single phonon with $\lambda = 2$ will give rise to a state with $J^{\pi} = 2^{+}$ and a state with 2 phonons of $\lambda = 2$ vibration will produce states with $J^{\pi} = 0^{+}, 2^{+}$, and 4^{+}.

It can be shown that the vector sum of μ components associated with projection of λ on the axis of symmetry will not produce an odd value of J. Parity of states is given as $(-1)^{\lambda}$.

Excited states with three $\lambda = 2$ phonons will produce states with $J^\pi = 0^+$, 2^+, 3^+, 4^+ and 6^+.

Hence, in nuclei with small surface deformation (δ), one can generate excited states based on the value of angular momentum of phonon.

4.21. Parameters Describing Ellipsoidally Deformed Nucleus

As mentioned earlier, the shape of a deformed nuclear surface is given as

$$R(\theta, \phi) = \overline{R}\left[1 + \sum_{\lambda\mu} \alpha_{\lambda\mu} Y_{\lambda\mu}(\theta, \phi)\right] \qquad (4.30)$$

where R = radius of a spherical nondeformed nucleus, $\alpha_{\lambda\mu}$ are deformation parameters, Y_{lm} are spherical harmonics, and θ, z are polar angles with respect to symmetry axis. The μ has $(2\lambda +1)$ values, which are integral values $+\lambda...........-\lambda$. if λ is the mode of surface deformation.

For spherical nuclei, $\lambda = 0$ and for quadrupole deformation lowest mode of deformation is $\lambda = 2$ and $\mu = -2, -1, 0, -1$, and -2 corresponding to five independent modes representing ellipsoidal shapes.

For octupole deformation $\lambda = 3$ and for higher values of λ, vibration states are also observed in such nuclei.

Kin etic energy of states associated with vibration of surface is given as

$$E_{KE} = \frac{1}{2}\sum_{\lambda\mu} B_\lambda \left|\dot{\alpha}_{\lambda\mu}\right|^2 \qquad (4.31)$$

where B_λ is a coefficient expressed in terms of mass density ρ and equilibrium radius R. Its value is given as $B_\lambda = \dfrac{\rho R^5}{\lambda}$.

The potential energy associated with vibration of surface is

$$U = PE = \frac{1}{2}\sum_{\lambda\mu} C_\lambda \left|\alpha_{\lambda\mu}\right|^2 \qquad (4.32)$$

where C_λ are deformation coefficients whose value depends upon the surface tension and Coulomb energy and is given as

$$C_\lambda = sR_0^2 (\lambda-1)(\lambda+2) - \frac{3}{2\pi} \frac{(Ze)^2}{R} \left(\frac{\lambda-1}{2\lambda+1} \right),$$

where S is the surface tension.

Thus, the total energy of a vibrating surface is given as

$$E_T = KE + PE$$
$$= \frac{1}{2} \left[\sum_{\lambda\mu} B_\lambda |\alpha_{\lambda\mu}|^2 + \sum_{\lambda\mu} C_\lambda |\alpha_{\lambda\mu}|^2 \right] \tag{4.33}$$

The oscillations are described by a set of harmonic oscillators with frequencies w_λ where w_λ is given as

$$w_\lambda = \left(\frac{C_\lambda}{B_\lambda} \right)^{\frac{1}{2}} \text{ where } C \text{ and } B \text{ are defined above.}$$

Energy eigenvalues of the harmonic oscillators are given as

$$E = \sum_\lambda \left(n_\lambda + \frac{1}{2} \right) \hbar w_\lambda$$
$$\tag{4.34}$$

where n_λ is the number of oscillations or phonons in the λ-mode of vibration.

After substituting values of C and B, one gets the value of w as

$$w_\lambda = \left[\frac{SR_0^2 (\lambda-1)(\lambda+2)\lambda}{\rho R_0^5} \right]^{\frac{1}{2}} \tag{4.35}$$

For $\lambda = 2$, $\lambda = 3$, $\lambda = 4$, angular frequency w is given in units of $s / \rho R_0^3$ as shown in the following:

$$w_{\lambda=2} = (const)\left((1)(4)2\right)^{\frac{1}{2}} \tag{4.36}$$

$$w_{\lambda=3} = (const)\left((2)(5)3\right)^{\frac{1}{2}} \tag{4.37}$$

one sees that

$$w_3 = \left(\frac{5.47}{2.82}\right)w_2 = 2w_2 \tag{4.38}$$

With very few exceptions, the first excited state in even-even nuclei have $J^\pi = 2^+$. The next higher excited state is 2 phonon states. The coupling of two phonons can give $J^\pi = 0^+, 2^+, 4^+$ and energies as $2w_2$, $4w_4$, etc.

The first excited state with one octupole phonon $\lambda = 3$ will be $J^\pi = 3^-$, and higher excited states would be due to vector coupling of 2 octupole phonons as $J^\pi = 3 \pm 3 = 0,1,2,3,4,5,6$, and parity will be $(-1)^{3+3} = +ve$ and energy $E_\lambda = \sum_\lambda n_\lambda \hbar w_\lambda$. Figure 4.2 due to Eichler (4.15) shows energies of 1st 2^+ excited state with $\lambda = 2$ phonon and 2nd excited state formed by the coupling of two $\lambda = 2$ phonons with states $J = 0$, 2, and 4, etc., and positive parities as shown in figure 4.12.

On the left side of the figure 3 states are degenerate but are separated in energies as shown in the figure 4.12b. Figure 4.12c shows the octupole state for $\lambda = 3$ and hexadecupole states with $\lambda = 4$ phonon.

Figure 4.13 shows vibration excited states of other even-even spherical nuclei in the medium mass region such as Cd^{114}, Sn^{118}, Te^{122}, and Ba^{134}. In each case, the first excited sate is 2^+, and the second excited state are triplets with 0, 2, and 4, and positive parities. Energies are given by the formula (4.32). The degenerate states have small spread in energies but the mean value is twice the energy of the 1st excited state.

Similarly, vibration states are also observed in some heavy nuclei such as Pt [192-196] and Hg [196-200].

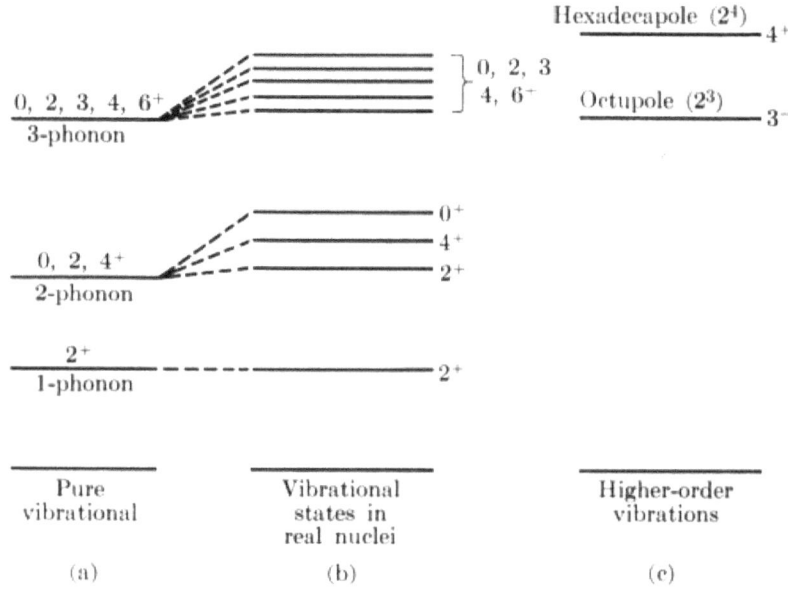

Figure 4.12 Schematic vibrational energy-level diagrams of medium-mass even-even nuclei. [From E. Eichler, *Rev. Mod. Phys.* **36,** 809 (1964).]

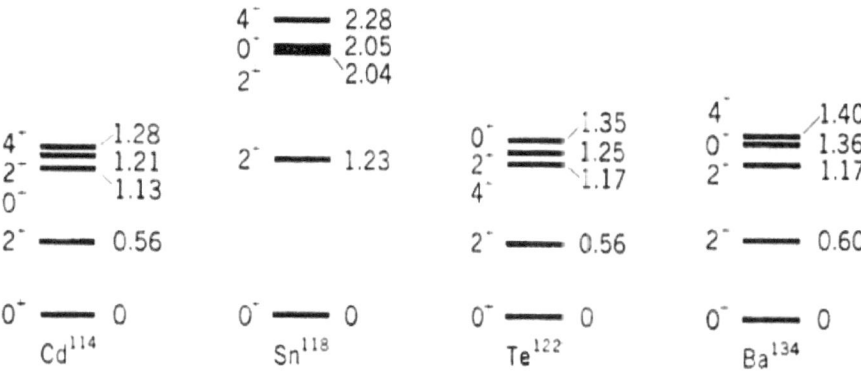

Figure 4.13 The lowest-energy states of various spherical even-even nuclei. Excitation energies (in MeV) are given at the right, and I^π are shown at the left. Note that in all cases the ground state is 0^+, the first excited state is 2^+, and there are three states at about twice its excitation energy with $I^\pi = 0^+$, 2^+, and 4^+, not necessarily in that order. In some of these nuclei, all three of the latter group are not yet known experimentally.

The vibration excitations are also characterized by enhanced electric transition probabilities.

These can be expressed directly in terms of parameters B_λ and C_λ. For one phonon, excitation transition probabilities are given as

$$B\left(E_\lambda, 0^+ \rightarrow 2^+\right) = (2\lambda + 1)\left[\frac{3}{4\pi} ZeR_0^\lambda\right]^2 \frac{\hbar}{2\left(B_\lambda C_\lambda\right)} \tag{4.39}$$

It is observed experimentally that the transition probabilities for many nuclei given as $B(E2, 0^+ \rightarrow 2^+)$ transitions are 10 to 50 times the single particle estimates. From the observed energy and $B(E2)$ value for excitation of the first 2^+ state, one can determine the parameters $B_{\lambda=2}, C_{\lambda=2}$ by using the expression above (4.32) and (4.33) as

$$\hbar w_\lambda = \hbar \left(\frac{C_\lambda}{B_\lambda}\right)^{\frac{1}{2}} \tag{4.40}$$

4.22. Unified Model or Nillson's Model

We had discussed earlier on page 84 that the shell model of nucleons had a great success in predicting the properties of ground states and low lying excited states of light mass nuclei. The collective vibration model had great success in predicting the properties of excited states of medium-weight nuclei, and the rotational model had great success in predicting the properties of deformed rare earth and heavyweight nuclei.

S. G. Nillson (4.16) developed a unified model for single particle nucleon motion in deformed potential. His theory is based on a harmonic oscillator well that is used for spherical nuclei with modifications for deformed well. This model is known as unified model or Nilsson's model.

According to this model, one uses a series of harmonic oscillator wave functions of the Hamiltonian operator.

Hamiltonian (H) of the nucleus in the deformed potential well can be written as

$$H = H_0 + H_\delta + C\,\vec{l}.\vec{s} + D\,\vec{l}.\vec{l} \tag{4.41}$$

where H_δ is the Hamiltonian for the single particle moving in a deformed potential well and is given as

$$H_\delta = \frac{4}{3}\sqrt{\frac{\pi}{5}}\delta m w_0^2 r^2 Y_{2,0}(\theta,\phi) \tag{4.42}$$

where δ is the deformation parameter and w is the angular frequency of the oscillator, and H_0 is the Hamiltonian for the spherical nucleus, and eigenvalues for H_0 are given as

$$E_0 = \left(N+\frac{3}{2}\right)\hbar w_0 \tag{4.43}$$

where N is the principal quantum number. Constants C and D are associated with spin-orbit and $l.l$ potential. D $l.l$ is a term used by Nillson as a correction term. It has the effect of distorting the harmonic-oscillator potential to a more flat-bottomed potential, closer in shape to the actual nuclear potential. For a deformed nucleus, these terms are negligible, and they can be omitted.

Nilsson's treatment for a deformed nucleus is as follows:

As has been discussed before that when a nucleus is deformed taking the shape of a prolate or an oblate spheroid, its two axes are equal, and the third unequal axis Z is the symmetry axis. Total angular momentum J and its projection along the Z axis are no longer conserved quantity, but projection of j along the symmetry axis Z known as Ω is the conserved quantity.

A total wave function ψ for a given set of the principal quantum number N and the angular momentum projection Ω of J along Z-axis in the distorted potential well is written as sum of wave functions as

$$\begin{aligned} \psi_\Omega &= \sum_R C_k \psi_k^0 \\ &= c_1\psi_1^0 + c_2\psi_2^0 + c_3\psi_3^0 \end{aligned} \tag{4.44}$$

where ψ_k^0 are the set of harmonic oscillator wave functions with given N, Ω, and k is a running index (not a quantum number). The objective of the calculations is that one has to find the coefficients (c_k) and the energy value for each of the final wave functions.

Nilsson's wave function ψ_k quantized along Z-axis can be specified by quantum numbers as $|N,l,\wedge\Sigma>$
where

N is the principal quantum number,

l = particle's orbital quantum number,

$\wedge$ = projection of l about the axis of symmetry,

Σ = projection of spin quantum number (s) about the axis of symmetry, which can take values $+1/2$ and $-1/2$.

Take the case of a single particle in $N = 2$, (s-d) shell, $N = 1$ is used for p-shell, and $N = 2$ for s-d shell.

Table 4.2 lists all harmonic oscillator wave functions for $N = 2$. All these functions are doubly degenerate, i.e., they can take two neutrons or two protons with opposite values of spin. Thus, an odd nucleon in s-d shell can give rise to states with $\Omega = 1/2, 3/2, 5/2$, where $\Omega = \wedge \pm \Sigma$, for $l = 2$ can take values 2, 1, and 0, and Σ can take values $\pm 1/2$. The $N = 2$ shell and $d_{5/2}$ orbit can accommodate $6 \times 2 = 12$ protons or 12 neutrons. Distortions expressed by H_δ mixes the states. Terms $Cl.s$ and $Dl.l$ give shell model wave functions and can be ignored for deformed nuclei.

Table 4.2. Harmonic oscillator wave functions for N=2, s-d shell

l/Ω	1/2	3/2	5/2
0	200+>		
2	220+>	221+>	222+>
2	221->	222->	

The wave function for deformed nucleus is written in terms of all harmonic oscillator wave functions in a given column.

Taking the example of nucleus O^{17}, which has the odd neutron in $d_{5/2}$ orbit. The wave function for Ω is given as the sum of harmonic oscillator wave functions as in the following:

$$\psi_\Omega\left(O^{17}\right) = a_1\,|200+> + a_2\,|220+> + a_3\,|221->\qquad(4.45)$$

where a_1, a_2, and a_3 are coefficients of wave functions in that state. Similarly for $\Omega_{\frac{3}{2}}$ and $\Omega_{\frac{5}{2}}$ states, one has

$$\psi_{\Omega=\frac{3}{2}}\left(O^{17}\right) = a_1 \mid 221+ > + a_2 \mid 220- > \tag{4.46}$$

$$\psi_{\Omega=\frac{5}{2}}\left(O^{17}\right) = a_1 \mid 222+ > \tag{4.47}$$

Values of coefficients a_1, a_2, a_3 for different values of δ have been calculated by Nilsson for many nuclei all across the periodic table. One can thus calculate the energies of states for $\Omega_{\frac{1}{2}}$, $\Omega_{\frac{3}{2}}$, and $\Omega_{\frac{5}{2}}$. Using the above wave functions for values of δ, the calculations made are shown in figure 4.14. Positive sign of δ is for prolate deformation, and negative sign is for oblate deformation.

Figure 4.14 is taken from Mottelson and Nilsson (4.17). One observes that each nucleon in a given shell-model orbit has different energies as a function of Ω values and for different values of δ.

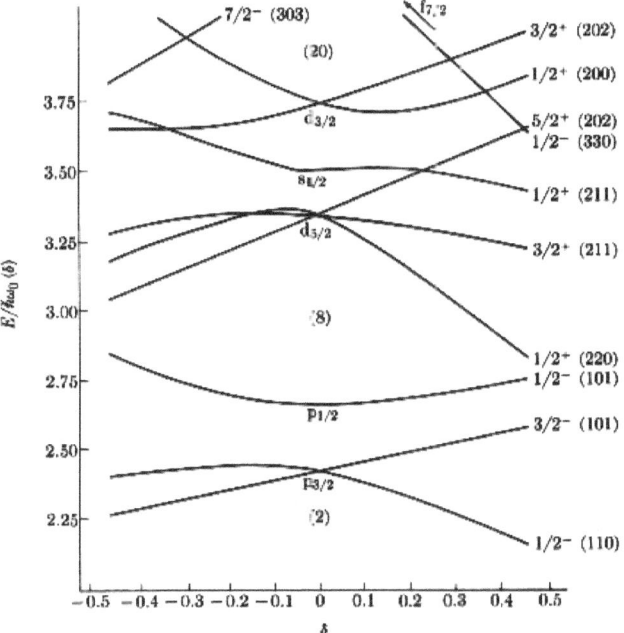

Figure 4.14 Energy of some low single-particle levels plotted vs. distortion parameter δ. [Adapted from B. R. Mottelson and S. G. Nilsson, *Mat. Fys. Skr. Dan. Vid. Selsk.* 1, No. 8 (1959).]

Various levels are identified by a half integral numerical indicating parity and three other numerals in parentheses. These are called the asymptotic quantum numbers $(N, n_z, \wedge)$, which are appropriate quantum numbers for an infinitely distorted axially symmetric harmonic oscillator potential.

N = principal quantum number

$\wedge$ = projection of l along Z axis

$n_z = n_3$ representsts the wave function of a harmonic oscillator along the third axis or axis of symmetry and is a constant of motion.

A single nucleon in $d_{\frac{5}{2}}$ state can give rise to 3 states with $\Omega = 1/2+$, 3/2+, and 5/2+, and their energies depend upon deformation δ-value. A nucleon in state $d_{3/2}$ will give rise to states with $\Omega = 1/2+$ and 3/2+, and a nucleon in state $S_{1/2}$ will give rise to $\Omega = 1/2+$ state. In general, for prolate deformation, the lowest energy state is lowest value of Ω, and for oblate deformation, the lowest energy state is for maximum value of Ω.

It is interesting to note that predicted values of J for $d_{5/2}$ state due to shell model is 5/2 whereas for prolate deformed nuclei, the lowest energy state is 1/2. Experimentally, J of F^{19}, which has a single proton in $d_{5/2}$ orbit is observed to have $J = 1/2$, which is consistent with Nilsson's model. Similar result is obtained for $_{10}Ne^{19}$, which has $J = 1/2$ instead of $J = 5/2$.

4.23. Liquid Drop Model

After the discovery of nuclear fission, Bohr and Wheeler (4.18) suggested a liquid drop model for the nucleus whereas the nucleus as a drop of liquid will gradually get deformed and then break into fragments produced in the fission.

This model was used by Von Weizsacker (4.19) to calculate the binding energies of nuclei. For this purpose, he proposed a semiempirical formula consisting of various energy contributions to total energy.

Volume Energy

These contributions come from volume energy, which arises from the neutral interactions of nucleons under the influence of nuclear force. It is found that the binding energy per nucleon becomes constant at about 8 MeV/nucleon over the entire mass region. Contribution from volume energy is proportional to mass atomic mass A. It is given as

$$B_1 = a_v r_o A^{1/3}$$

Surface Energy

This arises from the interaction of nucleons near or at the surface. Since the nuclear density decreases at or near the surface, there are fewer nucleon-nucleon interaction. Number of nucleons near the surface is proportional to surface area of the nucleus. As a result, this contribution is given as

$$B_2 = \alpha_s Area = \alpha_s \left(4\pi R^2\right) \tag{4.48}$$

where α_s is a constant whose sign is negative.

Coulomb Energy

Electrostatic repulsion force between protons has a long range characteristic, and the contribution is proportional to $1/r^2$. It also depends upon the distribution of protons in the nucleus. The potential energy contribution due to uniform distribution of charge over a surface of radius R is given as

$$B_3 = -a_c \left(\frac{Z_e}{R}\right)^2 = -a_c \frac{Ze^2}{r_0 A^{\frac{1}{3}}} \tag{4.49}$$

$$a_c = \frac{3e^2}{5r_0} \tag{4.50}$$

Asymmetry Energy (Neutron Excess)

Nuclear force is observed to be charge independent. That is, force between n-n, p-p, or n-p is the same and is of short range. However, there exist a repulsive electrostatic force between protons. Therefore, the nucleus will become unstable if there are too many protons. Looking at the periodic table, nuclei up to $Z = 20$ are stable if $N = Z$ whereas heavier nuclei tend to have more neutrons than protons. The reason being that excess of neutrons gives greater attractive force to balance the electrostatic repulsive force of protons. This gives rise to another term B_4 as in the following:

$$B_4 = -a_a \frac{(A-2Z)^2}{A} \tag{4.51}$$

Pairing Energy

Nuclei with even Z and even N are found to be more stable than those with even Z and odd N or with odd Z and even N whereas odd-odd nuclei are least stable. An additional term is included to give contribution from strong binding of even-even nuclei. This is given as

$$B_5 = +\delta \quad even-even \ nuclei$$
$$= 0 \quad even-odd \ nuclei$$
$$= -\delta \quad odd-odd \ nuclei$$

Thus, the binding energy of nuclei can be determined by adding contribution for the above terms as

$$BF = B_1 + B_2 + B_3 + B_4 + B_5$$

The sum of coefficients is as follows:

$$a_v A + a_s A^{\frac{2}{3}} + a_c \frac{Z(Z-1)}{A^{\frac{1}{3}}} + a_a \frac{(A-2Z)^2}{A} + a_p A^{\frac{3}{4}} \tag{4.52}$$

A fit to the experimental data of binding energy gives the values of these coefficients as follows:

$$a_v = 14.1 \; Mev$$
$$a_s = 13.0 \; Mev$$
$$a_c = 0.59 \; Mev$$
$$a_a = 19 \; Mev$$
$$a_p = 33.5 \; Mev$$

4.24. Summary of Nuclear Models

We have discussed various models of nuclear structure as they apply to nuclei in different mass region. In general, for light nuclei containing few nucleons, the ground state and excited states of nuclei are due to single or few nucleons. Due to the symmetry of nucleon wave functions, motion of nucleons is such that they cancel each other and give rise to simple properties such as the ground state total angular momentum for even-even nuclei is zero, and excited states are few in such nuclei.

As more and more nucleons are added, nucleons behave in a collective manner, giving rise to nuclear surface vibration and rotation of the nucleus as a whole. The properties of such nuclei are rather simple, mostly described by mechanical motion of a deformed rigid body. Since it is not possible to see the interior of the nucleus, these ideas of nuclear structure are based upon one's intuition and on the understanding of nuclear properties. The shell model was conceived based upon the success of Bohr's atomic structure of shells. This model was able to explain the abundance of elements based upon the shell closure. The observation of large quadrupole moments in some nuclei gave Rainwater to propose the idea of deformed shape of nuclei due to their collective motion. Fission of certain nuclei breaking in pieces could only be explained on the basis of deformation of nuclei. In order to unify the shell and collective model, Nilsson proposed the unified model. The unified model essentially combines the properties of shell and collective models in a single treatment and then modifying the Hamiltonian of the wave functions as it applies to either shell model or collective model. This treatment of nuclei has been found to be very successful and is mostly applied in determining the properties of nuclei.

The real nuclei are much more complex to be treated by any of the models discussed herein since these nuclei display features of several models superimposed on them. Therefore, the agreement of the experimental data of energies of excited states and transition probabilities with the theoretical predictions based upon any of these models is quite often qualitative and not exact.

An examination of the periodic table of elements shows that the most stable element is doubly closed-shell nucleus $_{82}Pb^{208}$. Nuclei heavier than lead are unstable. They decay with varying lifetimes. The heavier is the element, the shorter is the lifetime. Scientists believe that there may be another island of stable nuclei when the doubly closed shell numbers are reached. This possibility has invoked great interest in producing these elements by bombarding heavy elements with heavy ions such as C^{12} and O^{16}, etc. Such studies require very high energy accelerators. Such studies are being pursued in many laboratories and provide valuable information about the nuclear structure and nuclear reactions.

Fusion of heavy nuclei gives rise to proton-rich and neutron-deficient nuclei, which are unstable and are prone to decay by various decay modes. Study of mode of decay provides useful information about radioactive decays.

Heavy ion reactions can also produce nuclei, which are significantly deformed with excited states with angular momentum in excess of 40 ℏ. Study of these states would be of interest.

A detailed discussion of heavy ion reactions is beyond the scope of this introductory textbook.

CHAPTER 5

RADIOACTIVITY

5.1. Radio activity and Decay of Unstable Nuclei

In 1896, Henri Becquerel (5.1) accidentally discovered radioactivity. He discovered that certain elements were emitting certain rays. Further investigation of radioactivity was pursued by many other scientists, namely, Piere and Madame Curie and Ernest Rutherford. Becquerel and Curies were awarded Nobel Prize for their work.

Later, scientists showed that radiations emitted from radioactive nuclei were α, β, and γ rays. The α-rays are helium nuclei, β-rays are electrons and positrons, and γ-rays are electromagnetic waves.

An isolated system unstable toward particle or photon emission will, on an average, remain in its state of elevated energy for a certain time and then give out the excess energy by emitting radiations by the above modes of decay.

When a radioactive nucleus (called parent) decays into another nucleus (called daughter), there is a change of proton and neutron numbers. The α decay changes these numbers by 2 neutrons and 2 protons whereas β decay changes either a proton by one or a neutron by one. Gamma rays known as photons are massless and are emitted from excited states of nuclei.

5.2. Law of Radioactive Decay

In general, nuclear decay occurs at a certain rate. If there are N_0 number of radioactive nuclei and dN is the number of nuclei, which have decayed in a certain time interval dt. It was found that dN depends upon the original number N_0 and the time interval dt. This law known as the exponential decay law is expressed as

$$N = N_0 \, e^{-\lambda t}$$

where λ is a decay constant. Rate of decay of nuclei is defined by a half-life, which is the time for half number of nuclei to have decayed. These lifetimes vary widely for different radioactive nuclei from a long time of about 10^{18} years to very short time of about 10^{-8} sec.

5.3. Lifetimes Associated with Decay Constant λ

We know that $\left(\dfrac{N}{N_0} \right) = e^{-\lambda t}$ (5.1)

where N_0 is the original number and N is the number of nuclei remaining after the decay. For half the number of decays, one has

$$\left(\frac{N}{N_o} \right) = \frac{1}{2} = e^{-\lambda t_{\frac{1}{2}}}$$

or $\lambda t_{\frac{1}{2}} = \ln \dfrac{1}{2} = 0.69$

$$\lambda = \frac{0.69}{t_{\frac{1}{2}}}$$

Hence, if one measures half-life ($t_{\frac{1}{2}}$) of the decay of a radioactive nuclide, one can determine value of λ known as the decay probability. The values of λ depends upon the modes of various types of decay and are thus important quantities to compare with the theoretical predictions.

5.4. Particle Detectors

As mentioned above, unstable nuclei decay by emission of α, β, and γ-rays. Hence, a detailed study of these processes of decay requires detectors to identify the radiations and to measure their energies accurately. Over the years from 1930 onward, many such devices have been developed for the detection of these particles.

In the early years, detectors such as ionization chambers, proportional counters, Geiger-Müller counter, and scintillators were used to detect these particles. These detectors were relatively slow and not very reliable for the measurement of energies of particles. Hence, these detectors are not currently much in use, and I will not discuss these types of detectors.

Detection of charged particles electrons, positrons, protons, deuterons, α particles, and neutral particle neutron and γ-rays requires different techniques of detection and measurement of their energies.

I will discuss each type of detector in the section dealing with each type of decay.

5.5. The α Particle Decay of Nuclei

In α decay, the nucleon number A changes by four units, that is, daughter nucleus mass is A-4, and its proton number Z changes by 2.

Scientists have identified four families of naturally occurring radioactive nuclides, one with $A = 4n$ where n is an integer. Others with $A = 4n+1$, $A = 4n + 2$, and $A = 4n + 3$.

Figure 5.1 shows the uranium-radium $A = 4n + 2$ series with arrows indicating direction of decays, their types of decays, and half-lives of each mode of decay. The end product of all these series of decay is an isotope of lead, which is a stable nucleus.

Nucleus U^{238} has a long half life of 4.5×10^{19} years since the elements in our solar system are presumed to have been formed 4.5 billion years ago, a large fraction of U^{238} is still present on the planet.

Two other series of naturally occurring radioactive elements have been known for many years. These are (4n) Th^{232} series and (4n+3) U^{235} series. Their decays are shown in the figure 5.1b. Each of these decay series starts with α decay followed by one or two β decays. After successive α and β decays, the final end product is the stable nucleus of Pb^{209} and Pb^{207} respectively.

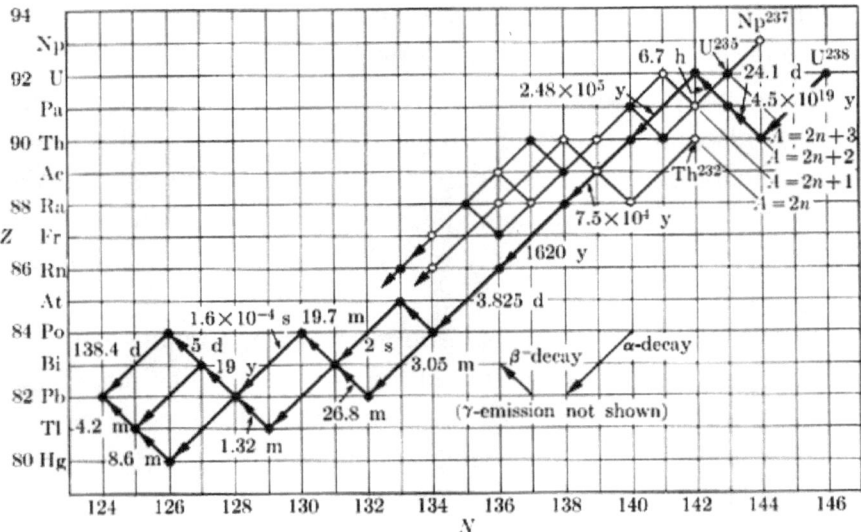

Figure 5.1. The $A = 4n + 2$ radioactive series and the first few links of the other three series. Where branching occurs, the half-life is written beside the main branch.

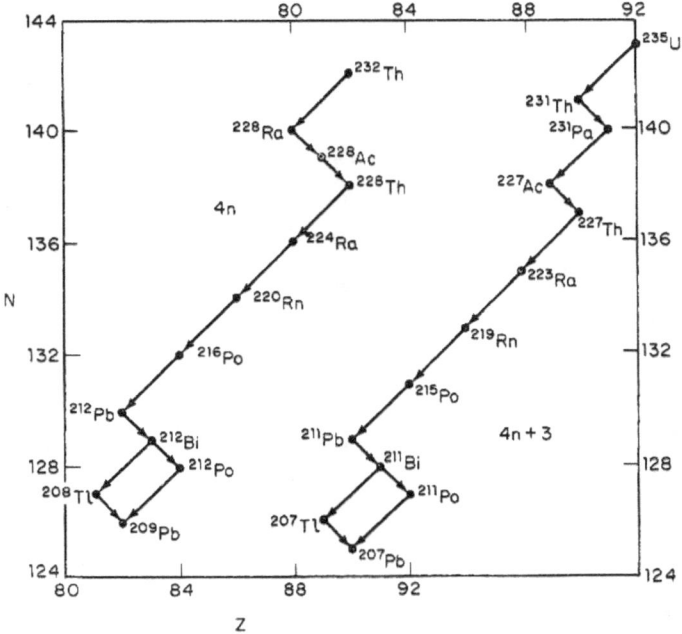

Figure 5.1b. Decay schemes of Th^{232} and U^{235}

One examines the reason under which an α particle involving (2p + 2n) decays in preference to decay of a single proton or a neutron. The plot of minimum and maximum binding energy for "last" nucleon and binding energies of "last" α particle in nucleus is shown in figure 5.2 for atomic masses from $A = 0$ to $A = 250$.

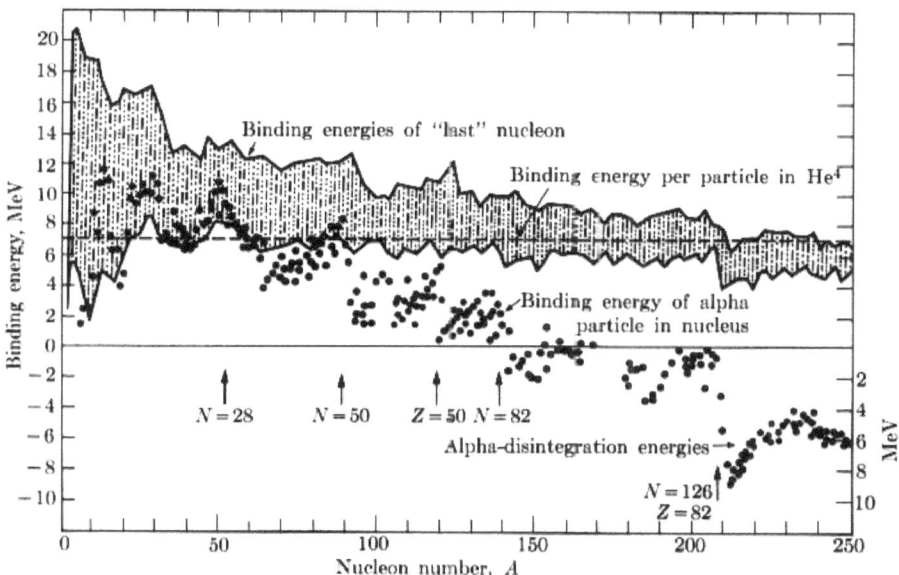

Figure 5.2. Minimum and maximum binding energy for "last" nucleon and binding energies of "last" alpha particle for beta-stable nuclei.

One notices that the binding energy of last nucleon decreases as a function of atomic mass A from about 2.0 MeV for very light nuclei to about 7 MeV for heavy nuclei.

The nuclei are therefore stable against breakup by nucleon whereas binding energy of last α particle in nucleus decreases rapidly and becomes less than zero for nuclei with mass $A > 150$, making such heavy nuclei unstable for α-decay.

Total binding energy of α particle is 28.3 MeV, meaning each nucleon is bound with energy of about 7.1 MeV per nucleon. Binding energy of each nucleon in a heavy nucleus is about 7 MeV. Thus, to remove four nucleons from inside the nucleus would require $7 \times 4 =$

28 MeV whereas when an α particle is formed, it would release 28.3 MeV thereby making the emission of α particle energetically possible for nuclei with mass $A > 150$. When the binding energy of last α particle in a nucleus falls below zero, the nucleus becomes unstable by emission of α particle.

Unit of radioactivity is a curie. One curie represents 3.7×10^{10} decays per second. Smaller units of millicuries and microcuries are commonly used in the measurement of radioactivity. Other units of radioactivity such as roentgen or rad are used in x-ray and medical field.

5.6. The α Particle Detector

The α particles are massive and carry two units of charge. Such particles, when they interact with matter, produce ionization of atoms due to Coulomb force produced by their electric charge. Such interaction imparts enough energy to the electrons and alpha particle loses energy.

Conservation of energy and momentum in a head-on collision of heavy particle of mass M with a light electron of mass m produces a loss of kinetic energy of few keV. A 5.0 MeV α particle will lose about 2 keV energy on the average for each collision. By successive collisions with other electrons, α particle loses all its energy and eventually stops in the detector.

In this way, one can determine the stopping range of α particles. This information is also important in determining the thickness of the detector for a given energy of α particle.

In the early years, α particle detectors used scintillators such as *ZnS* used by Rutherford to study α particle scattering. Nowadays, the most used detectors for measuring the energies of α particles are solid state detectors known as surface-barrier detectors and magnetic spectrometers.

5.7. Surface-barrier Detectors

The detector consists of a *p-n* junction of silicon and germanium crystals. Such a junction produces a valence band and a conduction band separated by an energy gap. When *n* type and *p* type materials are brought in contact, electrons from the *n* type material can diffuse across the junction into *p* type material and combine with holes. In the vicinity of the junction, the charge carriers are neutralized, creating a depletion layer.

The α particles entering the depletion layer creates electron-hole pairs. Electrons and protons flow in opposite directions to the applied voltage and give rise to a current producing an electric pulse. The detector is connected to a high voltage and is operated at low temperatures. Thickness of the depletion layer ranges from 10 micron to about 5 mm. A typical surface barrier detector for the measurement of energies of α particles may have a depletion layer of 100 microns and operates at 1,000 volts.

Such detectors have high energy resolution. Height of the electrical pulse generated is proportional to α particle energy.

5.8. Magnetic Spectrometers

The α particle energies are also measured by a magnetic spectrometer, which essentially measures its momentum. When α particles with many energies are emitted from a radioactive sources, these α particles pass through a large magnetic field produced by a 180° magnet. The source is placed at one end of the magnet, and a detector, usually a photographic plate, is located at the other end. The α particles experience a force perpendicular to its path from the magnetic field, which bends their path in an arc of a circle of radius *r* whose value depends upon their velocities. From the positions of their tracks in the photographic plate, one can determine their momentum.

A typical spectra of α particles emitted from the α decay of Th^{227} to the ground and excited states of Ra^{223} is shown in figure 5.3. The spectrum shows many peaks of resolved energies of α particles. These detectors have high energy resolution. Here, dE/E is about 0.1% whereas the resolution of solid state detectors is about 10 times worse.

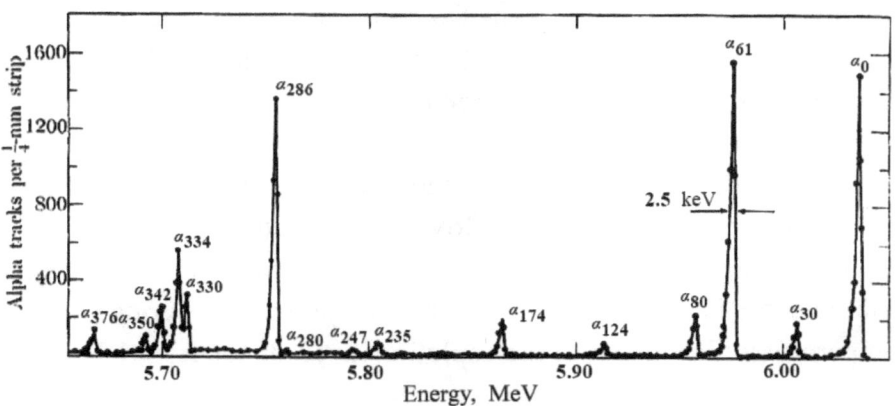

Figure 5.3. The α particle spectrum of Th^{227} taken by Pilger et al. with a magnetic spectrometer. The figure shows many α particle energies leading to the ground and excited states of the daughter nucleus Ra^{223}.

5.9. Experimental Decay Constant and Geiger-Nuttal Law

One measures the lifetimes of α particle decay. These lifetimes are observed to range from 0.3 μs (very short lifetime) to about 10^{17} years (very long lifetime).

Scientists, Geiger and Nuttal (5.2), observed a striking dependence of the half-life or decay constant (λ) on energies of the α particle emitted. They determined an empirical relationship between these two quantities given as

$$\log_{10} \lambda = C - D / \sqrt{E_\alpha} \tag{5.2}$$

where E_α = α particle energy measured in MeV, and C, D are slowly varying functions of Z (proton number) but are independent of the neutron number.

According to Geiger-Nuttal, if one plots $\log_{10}\lambda$ vs E_α (MeV) as shown in figure 5.3 below and if one finds value of λ for a given E_α, one can determine values of constants C and D.

This has been done for various observed α particle decay of even-even nuclei. As an example for a nucleus with $Z=90$, values of C = 52, D = 140 are determined. Using these values of constants C, D, and E_α one can determine from equation 5.2 the values of decay constant (λ) for a given Z value.

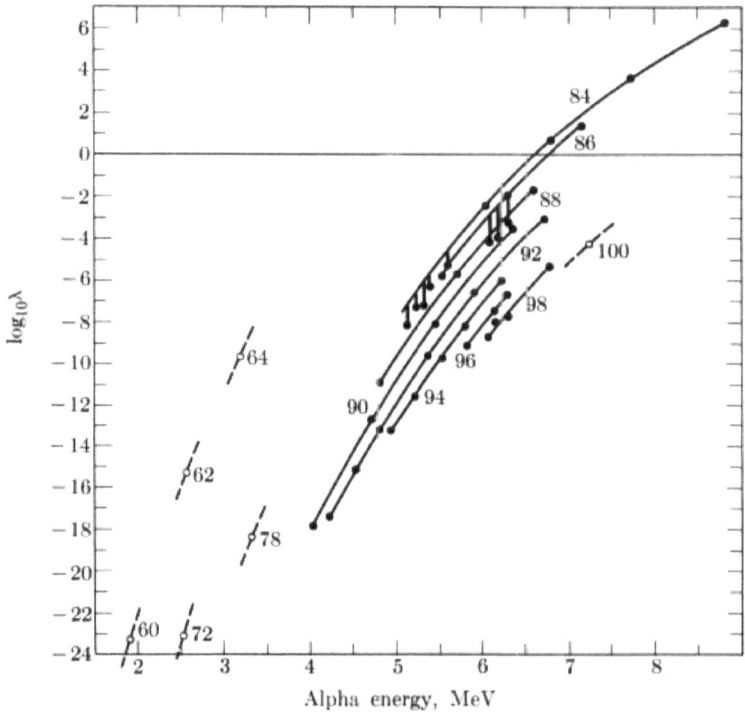

Figure 5.3 Logarithm of decay constant λ vs. alpha-particle energy for ground-state–to–ground-state decay of even-even nuclei.

5.10. Barrier Penetration Applied to α Particle Decay

The α-particle decay of a nucleus depends upon many factors. The probability of emission of α particles from inside a nucleus involves the penetration of potential barriers created by the Coulomb and nuclear forces. Transmission of charged particles through a potential barrier is a quantum mechanical phenomenon. Discussion of plane-wave transmission through different forms of potential barriers has been discussed in many books on quantum mechanics.

Rutherford, in studying α particle scattering from heavy nuclei, found that the Coulomb potential broke down at small distance from the center of the nucleus. Experimental data on α particle scattering from nuclei show that the scattering at large distances from the nucleus is given by Coulomb potential, and at around the nuclear surface, an additional attractive nuclear potential rapidly compensates the repulsive Coulomb potential.

Thus, a positively charged particle such as α particle is subject to a combination of short-range, attractive nuclear potential and a long-range, repulsive Coulomb potential. The Coulomb potential is $V = \dfrac{kZze^2}{r}$ where Z is the charge of the nucleus, $z = 2$, the charge of positively charged α particle and r (measured in fermi) is the distance from the center. From the nuclear surface outward, the potential seen by an α particle is repulsive and given as

$$V(r) = \frac{2.88Z}{r} - 1100 \exp\left[\frac{-r - 1.17A^{\frac{1}{3}}}{0.574}\right] Mev \tag{5.3}$$

where $V(r)$ is in MeV and r is in fermi.

This function is plotted in figure 5.4. The first term in the above equation is Coulomb potential. The second term is very small. The above equation is valid from nuclear surface and outward to infinity. Inside the nucleus, the alpha particle is bound to the nucleus and its constituents, 2 neutrons and 2 protons are orbiting. These four nucleons form the alpha particle at the nuclear surface, which tries to penetrate the potential barrier of the nucleus from inside to be emitted.

The potential barrier for α particles of energy 4.76MeV to be emitted from the nucleus $_{90}\text{Th}^{232}$ is about 20MeV.

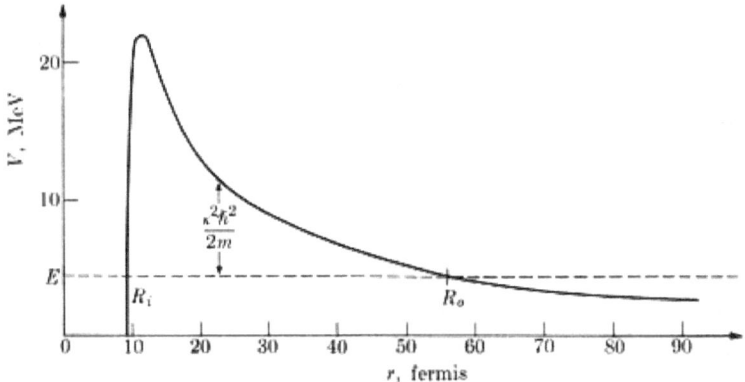

Figure (5.4)

Potential barrier between the nucleus $_{90}\text{Th}^{230}$ and an alpha particle as given by Eq. (5-13). The total energy of decay is $E = 4.76$ MeV.

The α particle energy in radioactive decay.

When a nucleus decays by a particle emission, a residual nucleus is left, which is known as the daughter nucleus (D). The energy released is shared between the alpha particle and the residual nucleus. Total energy (E) released in α decay is given by equation 5.4 where m_α and m_D are the masses of α particle and the daughter nucleus.

$$E = \frac{p_\alpha^2}{2m_\alpha} + \frac{p_0^2}{2m_D} = E_\alpha \left(1 + \frac{m_\alpha}{m_D} \right) \tag{5.4}$$

Since $m_\alpha = 4$, one can determine α particle energy as

$$E = E_\alpha \left(1 + \frac{4}{A} \right).$$

The energy released in the decay would be available to the α particle to escape from the nucleus. The probability of escape of α particle from the nucleus is given by λ, decay constant. Total decay constant (λ) is given as $\lambda = \lambda_0 B$, where B is the barrier penetration factor and λ_0 is the reduced decay constant in the absence of the barrier. The λ is a measured quantity whereas λ_0 can be estimated.

5.11. Theoretical Treatment of α Particle Decay

For a spherically symmetric potential, one can write the α particle wave function as a product of radial function and a spherical harmonic function

$$\psi = \left(\frac{u}{r} \right) Y_{l,m}(\theta, \phi) \tag{5.5}$$

and the radial wave equation is given as

$$\frac{d^2 u}{dr^2} + k_i^2 u = 0. \tag{5.6}$$

The wave number k inside the nucleus for $l > 0$ is given as

$$k_i^2 = \frac{2m}{\hbar^2}\left[E - V(r) - \frac{l(l+1)\hbar^2}{2mr^2}\right]$$ (5.7)

where $V(r)$ is given as in equation 5.3.

The l is the orbital angular momentum of emitted α-particle, and r is the nuclear radius in fermi; m is the reduced mass.

One can assume that the α particle impinging on the barrier from the inside can be represented by a radial wave function u_i^+, which contains $e^{-ik_i r}$ as a factor (plane wave with k_i as wave number).

Outside the barrier, one represents a radial wave function u_f^+, which contains $e^{ik_f r}$ as a factor. One defines (B) as penetration factor, which is the ratio of integrated flux outside the barrier to the integrated flux impinging on the barrier from inside. Integrating over the total sphere with $r^2 d\Omega$, one gets

$$B = \frac{k_f \int |\psi_f|^2 r_f^2}{k_i \int |\psi_i|^2 r_i^2} d\Omega$$ (5.8)

$$B = \frac{|u_f^2| k_f \int |Y_{l,m}(\theta,\phi)|^2 d\Omega}{|u_i^+|^2 k_i \int |Y_{l,m}(\theta,\phi)|^2 d\Omega} \quad \text{or } B = \frac{|u_f^2| k_f}{|u_i^+|^2 k_i}$$ (5.9)

After inserting the values of k_f, k_i etc , in equation (5.9), one obtains the value of B as

$$B = \exp\left(-2\int_{Ri}^{R_o} k\, dr\right)$$ (5.10)

where $k_i = \frac{(2m)^{\frac{1}{2}}}{\hbar}\left[V(r) + \frac{l(l+1)\hbar^2}{2mr^2} - E\right]^{\frac{1}{2}}$ (5.11)

$$k_f = \frac{\sqrt{2mE}}{\hbar}$$ (5.12)

Limits of integration is from R_i to R_0. These are known as turning points where $k = 0$ as shown in the figure 5.4

With numerical values inserted, one obtains a formula for k as

$$k = \frac{0.437}{(1+\frac{4}{A})^{1/2}} \left[\frac{2.88Z}{r} + \frac{5.23(1+\frac{4}{A})l(l+1)}{r^2} - 100\exp(\frac{r-1.17A^{\frac{1}{3}}}{0.574}) - E \right]^{1/2} \quad (5.13)$$

Equation 5.10 for B has been numerically calculated for a large number of cases with $l = 0$ in the mass region $60 \leq Z \leq 100$.

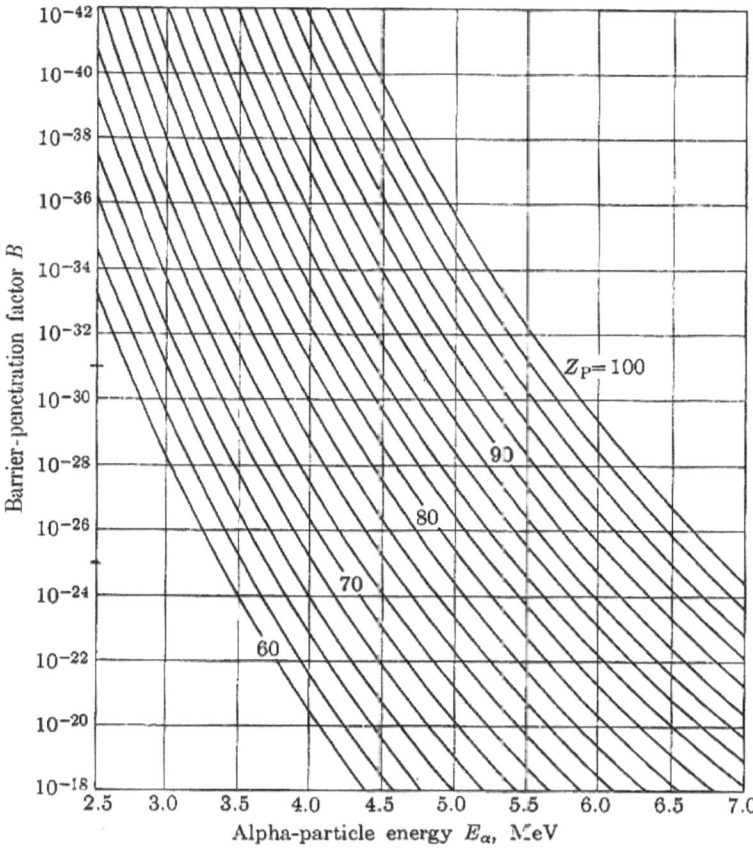

Figure 5.5. Barrier penetration factor (B) for $l = 0$, Z_p is the atomic number of the parent nucleus.

The results of these calculations are plotted in Figure 5.5 showing values of B as a function of E_α (alpha particle energy) for various nuclei of different Z-values.

As is seen from figure 5.5 barrier penetration factor B varies greatly by many orders of magnitude for a small change of about 1 MeV of α-particle energy.

The α particle with $l > 0$ is emitted between states of different J^π. Barrier penetration factor is not very sensitive to change in atomic mass A. For an increase in A of one unit, the change in B is about 4%. However, barrier penetration factor is quite sensitive to change in angular momentum carried by α particles.

For α particle emitted with $l = 0$, there is no change between the total angular momentum of the parent and the daughter nuclei. The lifetime of decay varies by many orders of magnitude for a change of few MeV in α particle energy. One can have α particle emission from a $0 \to 0$ state.

In table 5.2 below, one gives the experimental values of half-life, barrier penetration factor, reduced decay constant (λ_0), and reduced width (ΔE_0) for a number of nuclei in the mass region $144 < A < 254$.

In column 6 of this table are given the values in fermi of R_i—the distance between the center of mass of α particle for which kinetic energy is zero. Column 4 of this table gives the values of half-lives for α decays.

Reduced width is defined as

$$\Delta E_0 = \frac{\hbar}{\tau_{\frac{1}{2}}} \text{ where } \tau_{\frac{1}{2}} = \text{half-life and } t_{\frac{1}{2}} = 0.693 / \lambda_o .$$

In some cases, nuclear decay of α particle emission is followed by β particle emission. In the case where β emission competes with α emission, the decay constant of α decay is given as

$$\lambda_\alpha = \lambda - \lambda_\beta$$

where λ is the measured total probability and λ_β is the beta decay probability.

Table 5-2   Ground-state alpha transitions of even-even nuclei, $l = 0$

Atomic number Z	Mass number A	α-particle energy E_α, MeV	Particle half-life for α-decay, $t_{1/2}$, sec*	Ground-state group intensity, percent	Inner turning point R_i, fermis	Barrier penetration factor B*	Reduced decay constant λ_0, sec*	Reduced width ΔE_0, keV
60	144	1.90	1.58 (23)	100	8.44	2.18 (−42)	2.02 (18)	1.33
64	148	3.16	4.47 (9)	100	8.50	7.52 (−30)	2.06 (19)	13.6
72	174	2.50	9.5 (22)	100	8.77	5.44 (−43)	1.34 (19)	8.85
84	208	5.108	9.24 (7)	100	9.17	2.96 (−27)	2.54 (18)	1.67
84	212	8.780	3.04 (−7)	100	9.35	1.32 (−13)	1.73 (19)	11.4
84	218	5.996	1.827 (2)	100	9.32	1.31 (−22)	2.90 (19)	19.1
86	208	6.141	6.90 (3)	100	9.19	4.35 (−23)	2.31 (18)	1.52
86	218	7.130	1.90 (−2)	99.8	9.34	4.67 (−19)	7.82 (19)	51.5
88	224	5.680	3.15 (5)	94.8	9.34	5.91 (−26)	3.52 (19)	23.4
90	230	4.682	2.528 (12)	74	9.37	6.61 (−33)	3.07 (19)	20.3
92	234	4.768	7.83 (12)	72	9.40	2.34 (−33)	2.73 (19)	18.0
94	238	5.493	2.822 (9)	72	9.44	9.30 (−30)	1.90 (19)	12.5
96	242	6.110	1.404 (7)	73.7	9.49	2.16 (−27)	1.69 (19)	11.1
98	250	6.024	3.45 (8)	83	9.56	1.16 (−28)	1.44 (19)	9.5
100	254	7.200	1.150 (4)	83	9.62	4.09 (−24)	1.23 (19)	8.1

* The number in parentheses is the power of 10 by which the preceding number is to be multiplied.

Values of barrier penetration factor (B_l) for $l > 0$ and for a nucleus with $Z = 90$ and for $E_\alpha = 4.5$ MeV are given in table 5.1.

Table 5.1. Barrier penetration factor for $l > 0$

$l =$	0	1	2	3	4	5	6
$\dfrac{B_l}{B_0}$	1	0.84	0.60	0.36	0.18	0.078	0.028

It is quite obvious that the probability of α particle emission decreases as the angular momentum carried by α particle increases. The B_l decreases by about 20% for a change in one unit of l.

Figure 5.6 gives an example of α particle decay spectra from Th^{227} decaying to nucleus Ra^{223} taken by Stephens (5.3) and compiled by Fay Ajzenberg-Selov. The α particle energies are made with the help of magnetic spectrographs. Experimental studies of α particle decay for many heavy nuclei have been extensively studied.

In this figure, one sees that α decay consists of many α particle energies, each decay leading to either the ground state or to some of the excited states of the daughter nucleus. From these alpha particle energies given in column 2 of the figure, one determines the energies of the excited states in Ra^{223}. The last column gives the percentage of each type of decay mode.

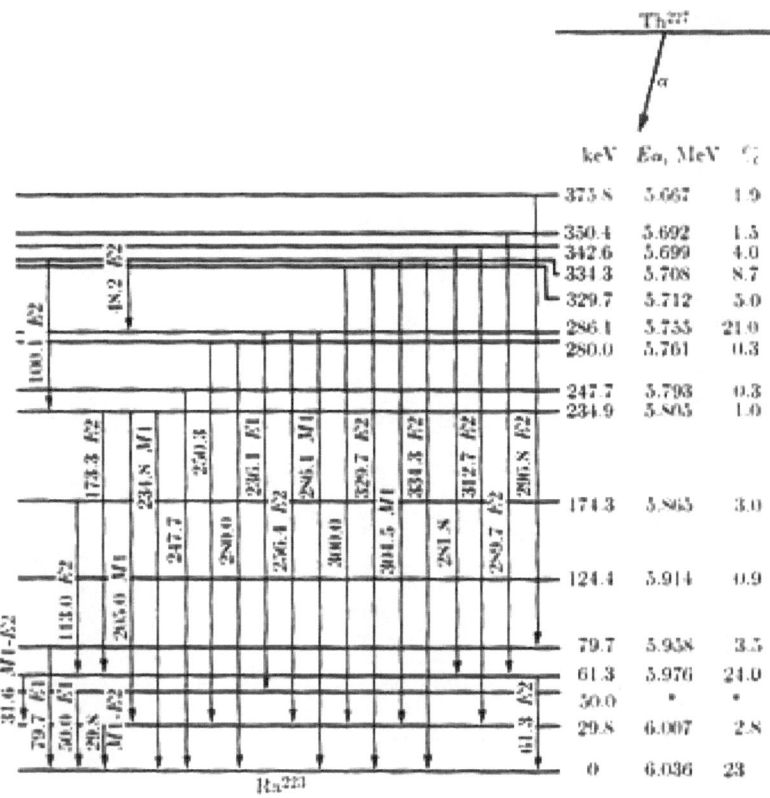

Figure 5.6. Energy level diagram of Ra^{223} obtained from α particle spectrum of Th^{227}. Also, it shows observed γ- ray transitions. (Taken from F. S. Stephens (5.3).

Labels E1, E2, M1, etc., represent the polarities of gamma rays emitted between excited states probability of decay to the ground state (*gs*) and the first excited state are very low. The reason being that these excited states built on *gs* band have positive parities.

Figure 5.7 shows α decay of Am^{241} to the ground state and excited states of Np^{237} due to Stephens (5.3). As is seen, 84.3% decay takes place to $\frac{5}{2}$ state, involving the angular momentum of alpha particle as $l_\alpha = 0$ or 2 from $\frac{5}{2}$ state of Am^{241}. These nuclei are known to be permanently deformed and give rise to rotational bands built on ground state and excited states. Thus, states built on the 5/2+ ground state of Np^{237} will have positive parities whereas the parity of Am^{241} is negative.

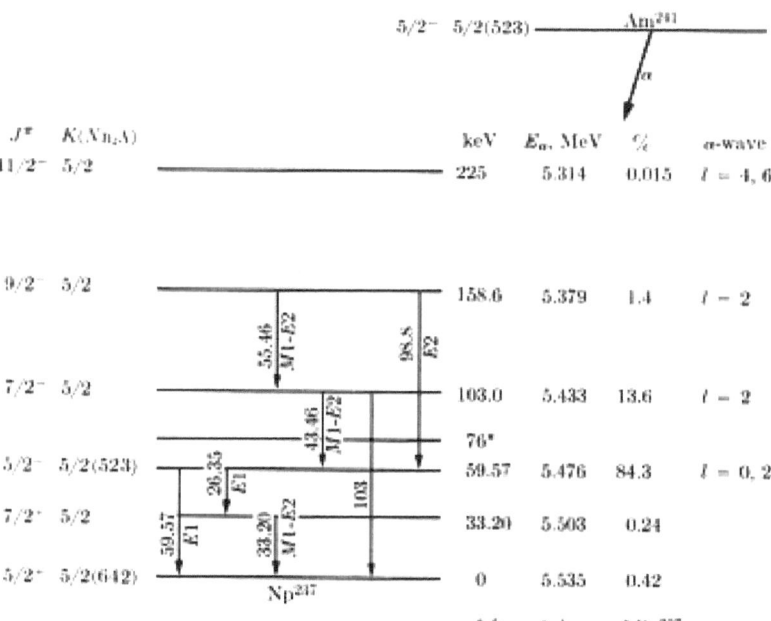

Figure (5.7) Decay scheme of Am^{241}, illustrating particularly the favored alpha decay to the 5/2− band. The l-values are the alpha waves believed to contribute to the population of each member of the favored rotational band. (From F. S. Stephens, op. cit., p. 198.

The decays from negative parity of Am^{241} to positive parity states of Np^{237} will be inhibited, and decay to negative parity states will be enhanced. Many nuclei when bombarded with nucleons such as neutrons or protons can emit α particles; an example is the reaction $F^{19}(p,\alpha)O^{16}$. This reaction was studied by Igo (5.4).

The Q-value of the above reaction is 8.118 MeV. Extensive studies of such reactions provide valuable information about the probability of α particle emission from nuclei. Such reactions will not be discussed further in this book.

5.12. Beta β Decay of Nuclei

An examination of table of nuclides reveals that among the nuclear species that are found in nature, there are many with identical A, which are stable isobars (nuclei having same mass). However, there are very few cases of naturally occurring stable neighbor isobars or pairs of types Sn_{50}^{115} and In_{49}^{115} where Z differs by one. In every case where such pairs exist, one member is actually unstable but may have an exceedingly long lifetime. In the above case, In_{49}^{115} decays to Sn^{115} with a half-life of $6 \times 10^{14}\ yrs$.

There is an abundance of stable isobars separated by two units of Z (and N). One observes that between two neighboring isobars, one with the larger atomic number will decay by a beta process to the lower one, but there is no process, which can change the nuclear charge by two units. Such a process is called double beta decay.

5.13. Detectors of Electrons and Positrons

Measurement of energies of β particles are important for the study of these decays. The measurement of end point energies of β decays provides information about the mass of neutrino as well as the mode of such decays.

In general, it is more practical to measure the momentum of these particles. A measurement of particle's momentum is accomplished by using magnetic spectrometers. The basic principle of such a device is deflection of electrons by a homogenous magnetic field of a magnet. When the electron having charge e and speed v enters the magnetic field B, it experiences a force that bends the electrons in a circular arc of radius r. The relation is given as

$$\text{Bev} = \text{mv}^2 / \text{r}$$

or Ber = mv = p, which is the momentum of the electrons.

The radioactive source is placed at one end of the spectrometer. Electrons pass through a collimator producing a fine beam. The electrons in passing through the magnet bend to the location of a detector and are detected by a solid state detector or a Geiger counter. Electrons of these energies have speed close to velocity of light. Measurement of radii of curvature of their paths in a magnetic field B provides β-rays momentum.

Electron energy is expressed in terms of its momentum and is given as

$$E_{kin} = [p^2 c^2 + (m_e c^2)]^{1/2} - m_e c^2$$

There are many variations of magnetic spectrometers. The basic requirement for any detector is the resolution $\Delta p/p$ and the transmission of particles.

One type uses a 180-degree chamber where the source is placed at one end, and the particle is detected at the other end of the chamber. This detector was discussed in section 5.8 for the measurement of energies of α particles.

Second type of spectrometer was designed by Starholm and Siegbahn and a prototype was built at Chalk River in Canada.

Figure 5.7b shows a portion of the electron conversion spectrum from the decay of Np^{239} taken by Evans and Graham (5.21).

The figure shows the intensity versus $B\rho$ in units gauss-cm. Lower figure shows improvement in resolution when the counter slit is reduced from 0.2 cm to 0.1 cm, but this reduces the transmission of electrons. The energy resolution is further improved when the counter slit width is 0.015 cm.

A typical beta decay energy spectrum of beta particle from Bi^{210} taken by Neary (5.6) is shown in figure 5.8.

This figure shows the energy spectrum of beta rays. The figure shows continuous energies of electrons from zero to maximum energy produced in the decay. This maximum energy is known as end-point energy.

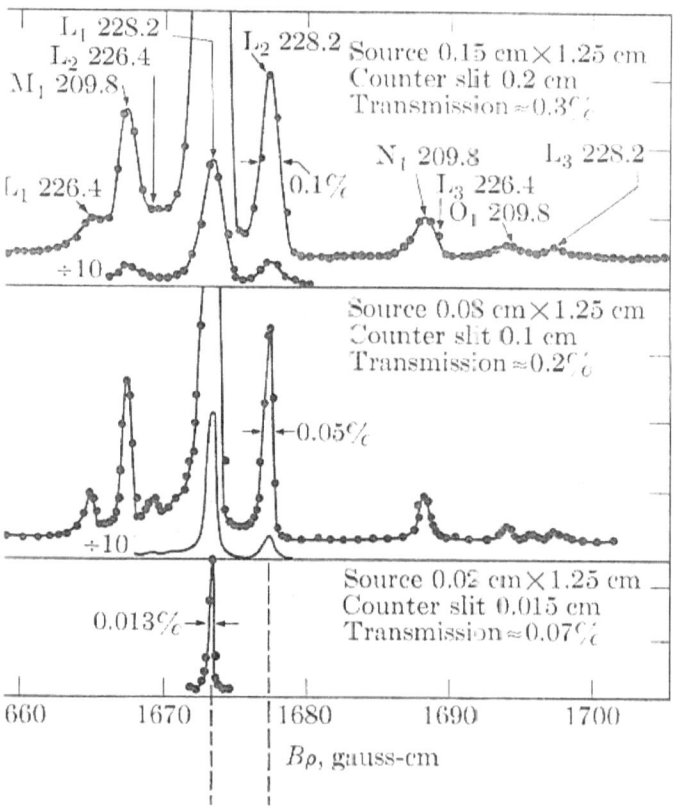

Figure 5.7b. Conversion electron spectra taken by a magnetic spectrometer

These data presented a serious problem of conservation of energy and momentum in beta decay. In 931, Pauli (5.7), in an effort to explain the violation of energy and momentum conservation, postulated the existence of a new particle, which he named as neutrino. Neutrino has an antiparticle known as antineutrino, which differs from the neutrino in terms of its direction of rotation.

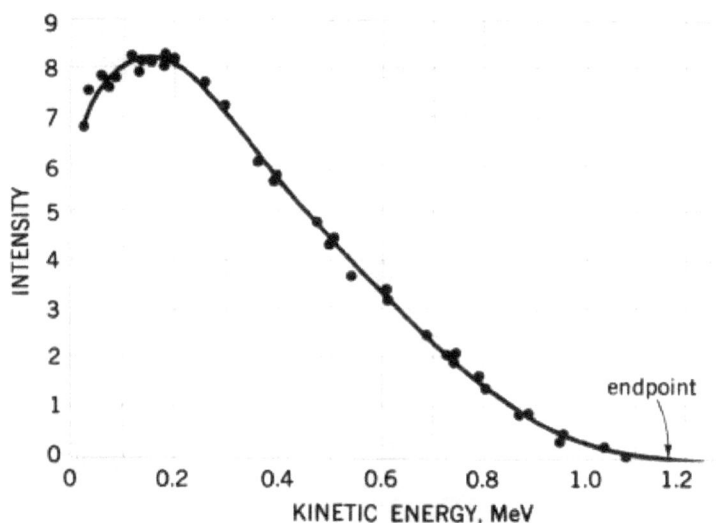

Figure 5.8. Energy spectrum of electrons emitted in the beta decay of Bi^{210}. [*From G. J. Neary, Proc. Phys. Soc. (London), A175:71 (1940).*]

5.14. Neutrino Properties

From the consideration of conservation of energy, charge, and spin, it was assumed that this new particle, neutrino, will have no mass, no charge, and spin = ½.

Thus, except for the spin, the neutrino will be similar to the photon and will be emitted near the speed of light.

Since such a particle does not interact with matter, it became impossible to detect its existence for a long time after its prediction.

Later as evidence gathered, another type of neutrino was postulated in the decay of pi-meson. Later, another type of neutrino was proposed in the decay of tau-meson. In 1961, experiments at Brookhaven laboratory confirmed the existence of the two types of neutrinos, one type accompanied by electron emission and the other type emitted from the decay of μ meson.

Reines and Cowan (5.8) performed the inverse β decay reaction by using a huge flux of antineutrino produced in a reactor. The reaction was

$$P^+ + v = n + e^+$$

The proton target was a scintillator containing large amount of water in which CdCl was dissolved. The positron created in the reaction produced ionization in the scintillator at the time of its emission and produced an electronic pulse. After slowing down, the positron captured an atomic electron and annihilated, producing two 0.51 MeV gamma rays, which were detected in a delayed coincidence with the electron, confirming the existence of antineutrinos.

Another strong evidence for the existence of neutrino was its detection at the time of supernova explosion. Scientists had built huge water detectors under mines in America and Japan. In 1987, a supernova explosion was witnessed by astronomers. At the same time, scientists at these undermine labs detected several events in their detectors, confirming the existence of neutrinos.

Besides such interstellar events, the sun is the source of large number of neutrinos, which are continuously being directed toward our planet; nuclear reactors also produce huge flux of neutrinos and antineutrinos. For example, in the decay of tritium to helium nucleus, an electron is emitted accompanied by this new particle antineutrino $\bar{v}$ as is shown below.

$$H_1^3 \rightarrow He_2^3 + e^- + \bar{v}.$$

Considering that the mass of the neutrino is assumed to be zero, the maximum energy of the electron would be the Q-value of the decay, which is 18 keV for this beta decay.

The Q-value of decay by e^- in this decay is calculated as

$$Q_\beta = \left[X_Z^A - X_{Z+1}^A + m_e \right] c^2 \tag{5.14}$$

where X_Z^A and X_{Z-1}^A are atomic masses of parent and daughter nuclei. In this decay, the daughter nucleus has an extra electron. The mass of neutrino is taken to be zero.

5.15. β Decay Modes of Unstable Nuclei

The β decay embraces all modes of nuclear decay in which the atomic number Z of a nucleus changes by one unit while the mass number A remains constant.

In β^-decay, a negative electron is emitted from the nucleus. An example of such decay is decay of free neutron given below.

$$n^0 \rightarrow p^+ + e^- + \bar{\upsilon}$$

In 1920, Scientist P.A.M.Dirac (5.9) predicted the existence of antiparticle for each known particle and thus predicted the existence of positron with positive charge as the antiparticle of electron. Similarly, antineutrino was predicted as the antiparticle of neutrino.

In the above decay, a neutron decays to a proton, emitting a pair of electron and antineutrino. The emissions of two leptons are governed by the law of conservation of lepton number, which states that the sums of leptons plus antileptons are conserved in any nuclear reaction or decay.

In β decay, a positive electron (positron) is emitted from the nucleus, accompanied by a neutrino such as in the decay of a proton to a neutron.

$$p^+ \rightarrow n^0 + e^+ + \upsilon, \; e^+ \text{ is an antilepton, and } \upsilon \text{ is a lepton.}$$

In this decay, a proton inside a nucleus decays to a neutron with the emission of a positron and a neutrino. Free protons never decay. Another example of positron decay is given below

$$F_9^{16} \rightarrow O_8^{16} + e^+ + \upsilon$$

in which a proton inside the nucleus F^{16} decays to a neutron forming O^{16} with the emission of a positron and a neutrino.

Another competing process called electron capture can also occur as shown below

$$p^+ + e^- \rightarrow n^0 + \upsilon$$

where the captured electron comes from one of the inner shell of the atom as K, L-shell, etc.

All β decays are possible only when the binding energy of the daughter nucleus exceeds that of the parent nucleus.

The masses entering in the above equation are atomic masses in the ground state. Very often, β decay leaves the daughter nucleus in an excited state. Kinetic energy T of disintegration product is then

$$T = Q - E_x$$

where E_x is the energy of excited state of the residual nucleus.

An example is β decay of Au^{198} to Hg^{198} in which the β decay takes place to the ground state and to the first two excited states is shown in figure 5.8b.

The figure shows that the most prominent decay (98.6%) is to first excited state of daughter nucleus Hg^{198}. This excited state then emits a 0.412 MeV gamma ray to the ground state.

In β decay, as in other decay process, the most important parameter that can be experimentally determined is the half-life or decay constant (λ) and the energies of the emitted β rays. From such measurements, one can determine momentum distribution of β particles emitted in the decay.

These quantities then can be compared with theoretical predictions.

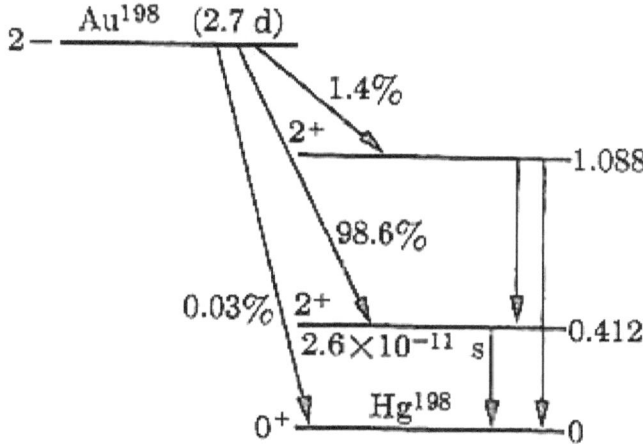

Figure 5.8b. Decay scheme of Au^{198} to the ground and first two excited states of Hg^{198}

5.16. Simple Theory of Beta Decay

In β decay, total energy available is divided between β particle, the neutrino, and the recoiling residual nucleus. Total energy available in the decay is the Q-value discussed above. The recoil energy is negligible, and all energy is divided between the beta particle and the neutrino and is given as $E_0 = E_e + E_v$, where E_0 is the maximum energy released in the β decay.

Since both leptons are emitted with velocities near velocity of light, their velocities are taken as equal to c.

Total energy of the neutrino (v) can be written as

$E_v^2 = \left(m_v c^2\right)^2 + \left(p_v c\right)^2$, where $m_v c^2$ is rest mass energy and p is the momentum of the neutrino; c is the velocity of light.

The fact that β ray spectra and that of neutrino is continuous indicates that final states available are continuous in energy. The probability of decay thus depends upon the density per unit energy or ($\dfrac{dN}{dE}$) of final states.

Figure 5.9 describes the overall picture of an initial state of an unstable decaying nucleus to a final state of the residual nucleus.

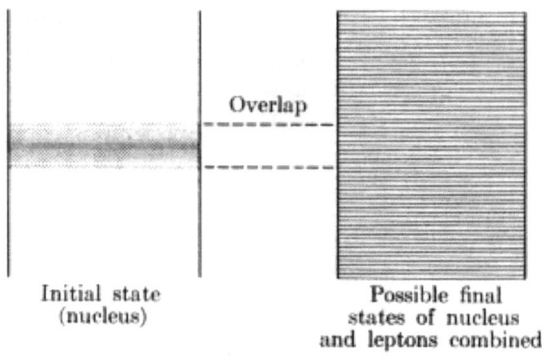

Initial state
(nucleus)

Possible final
states of nucleus
and leptons combined

The width of an unstable level overlaps a number of possible
levels of the nucleus-plus-lepton system.

Figure 5.9

Total width of the initial state of an unstable nucleus is shown on the left side of the above figure. This state overlaps with a number of possible states of the final residual nucleus plus two leptons.

The β decay of the parent nucleus can take place to any one of a large number of final states with slightly different energies without violation of law of conservation of energy.

From the usual quantum mechanical treatment of time-dependent perturbation, one can write the probability of decay between an initial state (i) and the final state (f) as

$$\lambda_{if} = \frac{2\pi}{\hbar} |M_{if}|^2 \frac{dN_f}{dE} \tag{5.15}$$

where $|M_{if}|$ is the matrix element for the transition from an initial to a final state and is given as

$$M_{if} = \int \psi_f^* H' \psi_i d\tau \tag{5.16}$$

where H' is the interaction energy operator causing the β decay.

This interaction of beta decay involves a new type of weak nuclear force. Theory of β decay has been developed extensively by Fermi (5.10), called Fermi interaction of β decay. The ψ_f is the wave function of the final state consisting of the decayed nucleus, β particle, and v. The ψ_i is the wave function of the initial state of the decaying nucleus, and dN/dE is the density of the final states.

The physical picture of this decay is that the parent nucleus in a given quantum mechanical state can change to a daughter nucleus in a given state plus two leptons (e, v) through the action of a perturbing potential. The two leptons can travel in any direction to form the nucleus, and they can share the available energy between them in a larger number of different ways.

The energy of the parent nucleus is not sharp so that it will overlap with a number of states of the daughter nucleus.

The matrix element M_{if} depends upon the H' where it is given as equation 5.16. One did not know the form of interaction operator H', but Fermi suggested trying the simplest possible operator namely a coupling constant G.

Final state of the total system is a product $\psi_{f_N}^* \, \psi_e^* \, \psi_v^*$. Hence, one can write H' as

$$H'_{if} = G \int \psi_{f_N}^* \, \psi_e^* \, \psi_v \, \psi_{iN}^* \, d\tau \qquad (5.17)$$

where ψ_{iN} is the normalized wave function of the parent nucleus, and $\psi_e \, \psi_v \, \psi_{fN}$ are normalized wave functions of electron and neutrino and daughter nucleus.

One can also show that by introducing V, volume of the nucleus, in which a large number of states with different values of angular momenta of leptons are contained. Assuming that wave functions of both leptons are treated as free particles contained in a volume V.

The outgoing electron and neutrino wave function with energy E is given as

$$\psi_e^*(r) \propto V^{-\frac{1}{2}} e^{ik_e r}$$

$$\psi_r^*(r) \propto V^{-\frac{1}{2}} e^{ik_r r}$$

where e^{ikr} represents a plane wave undistorted by the nuclear potential.

Integrating over the volume V, one gets

$$\int \psi_e^* \psi_r^* = \frac{1}{V} \int e^{ik_r r} e^{ik_e r} d\tau$$

$$\int e^{ik_e r} d\tau = 1 \quad \text{and} \quad \int_V e^{ik_r r} d\tau = 1 \qquad (5.18)$$

$$\int \psi_e^*(r) \psi_r^*(r) d\tau = \frac{1}{V} \times (\text{coulomb factor}) \qquad (5.19)$$

5.17. Coulomb Correction Factor

For charged particles such as e^- and e^+, emission of these particles from inside the nucleus is affected by the Coulomb force of the nucleus. Electron is attracted, and positron is repelled by the nucleus in order to make correction for the Coulomb effect. Fermi calculated the form factor, which is known as fermi function and is given as

$$F(Z, E_e) = \frac{2\pi\eta}{1 - \exp(-2\pi\eta)} \tag{5.20}$$

The above function depends upon Z and maximum energy of the electron where $\eta = \pm \dfrac{Ze^2}{4\pi\epsilon_0\, \hbar v}$.

The (+sign) is for e^- while (- sign) is for e^+. Equation 5.20 is based on nonrelativistic calculations. A plot of fermi function versus kinetic energy T of β particle for nuclei with $Z = 0\text{-}100$ prepared by National Bureau of Standards is shown in figure 5.10. Coulomb function enhances the probability for electron emission and retards the probability for positron emission. At high energies of beta particles, the Coulomb factor loses its effect on the spectrum shape. Hence, combining equations 5.17, 5.19 and 5.20, we get

$$\left|H_{if}'\right|^2 = \frac{G^2}{V^2}\left|M_{if}\right|^2 F(Z, E_e) \tag{5.21}$$

where M_{if} is the overlap integral of the final and initial wave functions of the nucleus given by equation 5.16 and

$$M_{if} = \int \psi_{fN}^* \psi_{iN}\, d\tau \ .$$

This nuclear matrix element can be computed in few cases where the structure of the nuclei is well known.

An example of such a decay is neutron decay

$$\left(n^0 \rightarrow p^+ + e^- + \bar{v}\right)$$

In this decay, M_{if} is equal to 1 since the wave functions of the neutron and protons are the same.

As discussed earlier, decay probability between an initial state (i) and final state (f) is given in (5.15).

Hence, for the determination of λ, one also has to determine the value of $\dfrac{dN}{dE}$, density of final states.

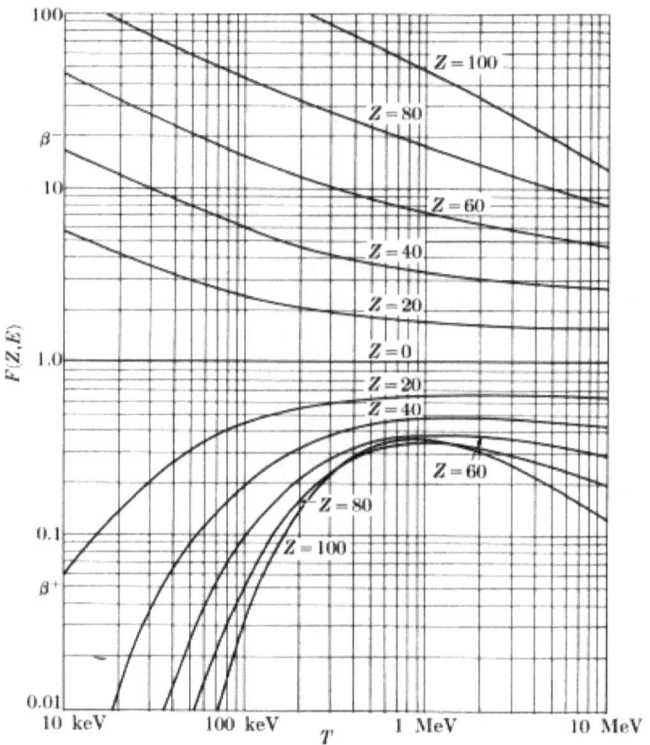

Figure 5.10   The Coulomb factor (Fermi function) $F(Z, E)$ plotted vs. T, the kinetic energy of the electron or positron. (Prepared from *Tables for the Analysis of Beta Spectra*, National Bureau of Standards, Applied Mathematics Series, 13.)

5.18. Measurement of Beta Particle Momentum

Decay usually involves study of β—the measurement of linear momentum of beta particles. This is done by using magnetic spectrometers. We know that a charged particle traversing the magnetic field gives rise to a circular motion of the particle with its radius or diameter proportional to its momentum.

Typical spectra of momentum of electrons and positrons in the beta decay of nuclei taken by Reitz (5.11) is shown in figure 5.11.

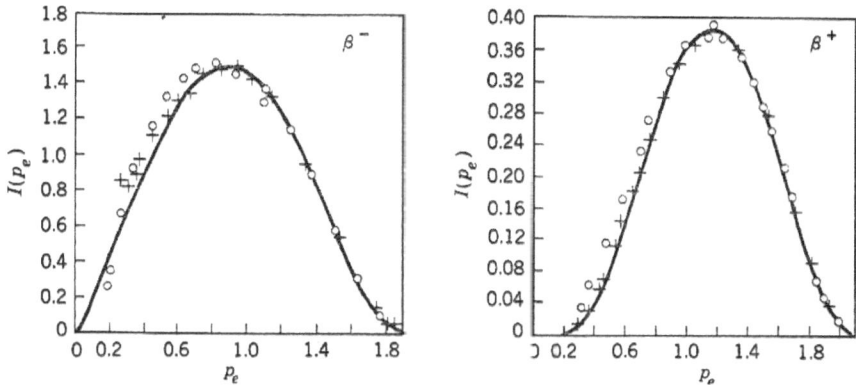

Figure 5.11. Momentum spectra of electron and positrons
from β- decay of $_{29}Cu^{64}$ to $_{30}Zn^{64}$ and from $\beta+$ decay of Cu^{64} to Ni^{64}.

The left figure is the electron momentum spectra for β decay of $_{29}Cu^{64}$ to $_{30}Zn^{64}$, and right spectra is β^{+} momentum distribution for beta decay of $_{29}Cu^{64}$ to $_{28}Ni^{64}$ particles versus momentum shows that these two plots of intensity of β momentum distribution is continuous and has an end point of maximum momentum determined by the Q-value of the β decay. Shapes of the two spectra differ due to the effect of Coulomb factor.

5.19. Calculation of Density of States of Residual Nucleus

Using classical approach of treating the behavior of particles contained in a cubical box of volume V of each side, L, the motion of leptons striking the sides of the box gives rise to the distribution of momentum of leptons. Thus, one can find the number of particles with momentum p and $p + dp$ inside the box. From these, one can determine the density of states, which is given as

$$dN = \frac{Vp^2 dp}{2\pi^2 \hbar^3} \text{ (where } V = \text{volume).} \tag{5.22}$$

Therefore, λ can be expressed in terms of momentum distributions of the β particle or neutrino. In the equation $P(p_e)dp_e = \lambda dN_e$, where P is the probability of electrons with momentum p and $p + dp$, and dN_e is the number of states with electrons having momentum p and $p + dp$.

Total energy E_0 available for decay is shared between an electron and neutrino given as $E_0 = E_e + E_v$.

For a fixed value of E_e, one can calculate the number of states with momentum of neutrino p_v and $p_v + dp_v$. The relativistic equation for energy of neutrino is the sum of rest mass energy and kinetic energy given as

$$E_v^2 = \left(m_v c^2\right)^2 + \left(p_v c\right)^2 \tag{5.23}$$

and $\qquad dp_v = \dfrac{E_v}{p_v c^2} dE_v = \dfrac{E_v}{p_v c^2} dE_0 \tag{5.24}$

After combining equation 5.22, 5.23, and 5.24, one obtains the density of neutrino states as

$$\frac{dN_v}{dE_0} = \frac{V}{2\pi^2 \hbar^3 c^3} \left(E_0 - E_e\right) \left[\left(E_0 - E_e\right)^2 - \left(m_v c^2\right)\right]^{\frac{1}{2}} \tag{5.25}$$

Probability of beta decay per unit time for decay into any of the electron states in the momentum interval dp_e is given as $P(p_e)dp_e = \lambda_k dN_e$ where λ is decay constant.

$$P(p_e)dp_e = \frac{G^2 \left|M_{if}\right|^2}{2\pi^2 \hbar^7 c^3} p_e^2 F\left(Z, E_e\right)\left(E - E_e\right)\left[\left(E_0 - E_e\right)^2 - \left(m_v c^2\right)^2\right]^{-\frac{1}{2}} dp_e \tag{5.26}$$

Substituting the values of H_{if} from 5.21 and density of states from 5.23 and assuming that $m_v = 0$, one gets

$$P(p_e)dp_e = \frac{G^2 \left|M_{if}\right|^2}{2\pi^2 \hbar^7 c^3} p_e^2 \left(E_0 - E\right)^2 F\left(Z, E\right) \tag{5.27}$$

The shape of beta spectrum is mainly determined by the factor

$$p_e^2 \quad \& \quad \left(E_o - E_e\right)^2$$

By taking the square root of equation 5.27, one gets

$$\sqrt{\frac{P(p_e)}{p_e^2}} = \frac{G\left|M_{if}\right|}{\sqrt{2\pi^3 \hbar^7 c^3}} \left(E_0 - E_e\right)\sqrt{F(Z, E_0)} \tag{5.28}$$

5.20. Kurie Plot, Measurement of β Particle Energies

If one plots square root of $P(p)/p^2$ on the left side versus energy E_0, provided the mass of neutrino is zero, one would obtain a straight line intersecting the energy axis with E_0, the maximum energy of β particles known as end point energy. However, if m_ν is not zero, Kurie plot would show a deviation from the straight line with end point energy less than the maximum energy by the amount of rest mass energy of the neutrino.

A large number of Kurie plots for β decays of unstable nuclei have been made, and the data had been analyzed to determine the end point energy of β particles. Various precise experiments have been carried out to determine the mass of neutrino from the Kurie plots. Any deviation from the end point energy can provide neutrino mass.

A typical Kurie plot (5.12) for beta decay of H^3 to He^3 taken by Langer et al. (5.13) is shown in figure 5.12.

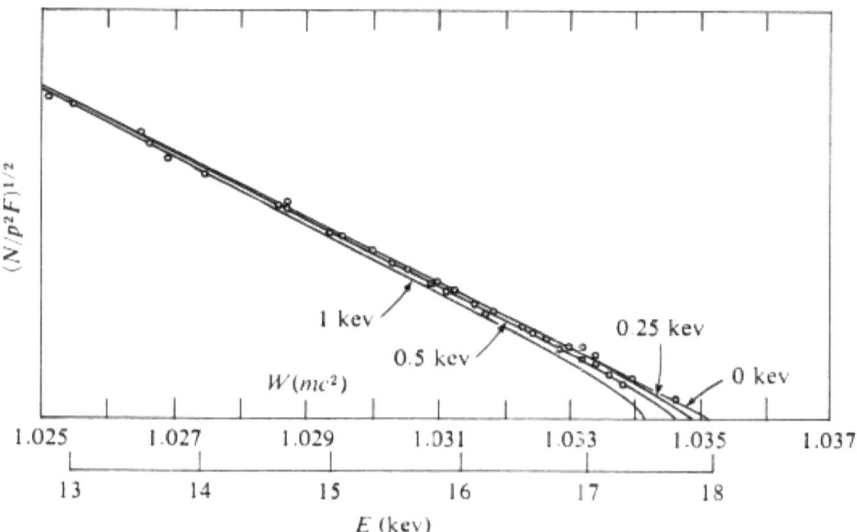

Figure 5.12. Comparison of the end portion of experimental points of H^3 Fermi plot with curves for different neutrino rest-masses. [From Langer, L. M., and R. D. Moffat, *Revs. Mod. Phys.*, 23, 10, (1951).]

This plot shows four lines drawn through experimental data based on the assumption made about the mass of the neutrino. One line assumes that the mass of neutrino is 1 keV, another assumes the mass to be 0.5 keV, and third assumes the mass to be 0.25 keV. The measurement of Kurie plot in the decay of $H^3 \rightarrow He^3$ with end point energy of 18.1 keV shows a deviation consistent with the mass of neutrino less than 0.25 keV/c^2, which is approximately equal to 5×10^{-4} m $_e$.

Equation 5.28 gives the form of momentum spectrum of the electrons emitted in an allowed beta decay. By comparing this formula with the experimental beta spectrum, one can

1.   test the theory of β-process,
2.   obtain information about whether the process is allowed or forbidden,
3.   obtain the value of Matrix element, $\left| M_{if} \right|$.

A straight line curve of Kurie plot provides support that the theory of β decay is essentially correct for allowed transitions. In general, Kurie plots for forbidden transitions deviate from straight line, and the data has to be corrected to make them straight.

5.21. Comparative Half-lives

By integrating the momentum spectrum from 0 to p_{max}, one can obtain the total decay probability (λ) as

$$\lambda = \frac{\ln 2}{t_{\frac{1}{2}}} = \int_0^{p_{max}} P(p)dp \qquad (5.29)$$

$$\lambda = \frac{m_0^5 G^2 c^4}{2\pi^3 \hbar} M_{if}^2 \int_0^{p_{max}} F(Z,E) \frac{p^2}{m_0^2 c^2} \frac{(E_0 - E)^2 dp}{m_0^2 c^4 m_0 c} \qquad (5.30)$$

Integral in equation 5.30 is a function of p, E_0 and Z and has been made nondimensional by dividing by $m_0^5 c^7$. One can simplify this integral by defining a quantity $f(Z, E_0)$ as in the following:

$$f(Z, E_0) = \int_0^{p_{max}} F(Z, E) \frac{p^2}{m_0^2 c^4} \frac{dp}{m_0 c} \qquad (5.31)$$

This relation has been evaluated by Feenberg and Trigg (5.14) for different values of Z and E_0 and are available elsewhere. The result is given as follows:

$$f(Z, E) t_{\frac{1}{2}} = \frac{2\pi^3 \ln 2 \, \hbar^7}{m_0^5 G^2 C^4 |M_{if}|^2} \qquad (5.32)$$

This is called comparative half-life also called ft values. Equation 5.32 shows that ft is inversely proportional to M_{if}; hence M_{if} can be determined from the measurement of $f(Z, E)$ and $t_{\frac{1}{2}}$ since values of $f(Z, E)$ and other constants are known.

In general, for β- transitions between $J = 0^+$ states of similar nuclear structure $|M_{if}|^2 = 2$. Knowing the value of M_{if} the ft values for those transitions can give value for coupling constant G. For example, in the β- decay of $O_8^{14} (\beta^+) N_7^{14}$, the following measurements of half-life and Q-value have been obtained as $t_{\frac{1}{2}} = 72.5$ s and $T = 1.810$ MeV. The fermi function and ft values are $f(Z, E) = 42.8$, $ft = 3103$ s, and $M_{if} = 2$.

By inserting these and constant values in the equation 5.32, one finds $G = 1.4 \times 10^{-62} \, Jm^3$ or by converting energy in MeV and distances in fm, one gets the value as $G = 0.9 \times 10^{-4} \, Mev \, fm^3$.

The force involved in β decays is much weaker than the nuclear force between nucleons.

According to modern theory, the weak interaction is caused by W^+, W^-, and Z elementary particles, which were discovered at the center of high energy nuclear research located in Geneva, Switzerland.

5.22. Super Allowed, Allowed, and Forbidden Transitions Rules

Super allowed decays have the shortest half-lives with log ft values of about 3.5.

A histogram obtained from nuclear data sheets showing number of β^-decay nuclei versus values of $\log_{10} ft_{\frac{1}{2}}$ is shown in figure 5.13.

This figure shows large number of allowed and super allowed decays as opposed to forbidden decays.

Decay probabilities or lifetimes of beta decay depends upon many factors such as energy of beta particles, the angular momentum carried by the two leptons, change of parity, and degree of overlap between the initial and final states wave functions.

Spins of electrons and neutrinos are known to be ½. That is, the two spins are $S_e = \dfrac{1}{2}$, $S_\upsilon = \dfrac{1}{2}$.

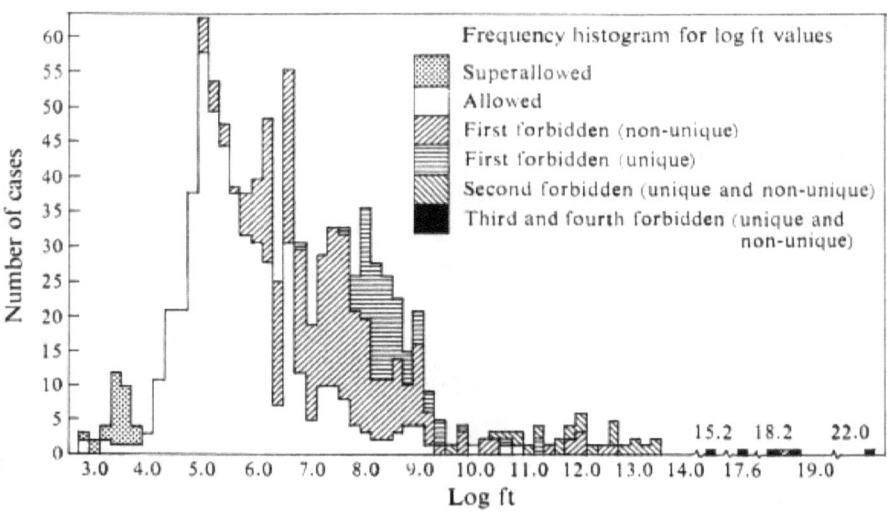

Figure 5.13. Histogram of the number of cases of nuclides for different log ft values. The cross-hatched area is for mirror nuclei transitions, and the dotted areas are the unique-shape transitions. [From Nuclear Data Sheets: 5-5-140 (Nov. 1963).]

These spins can orient either parallel giving a value of $S = \dfrac{1}{2} + \dfrac{1}{2} = 1$ or antiparallel giving a value of $S = \dfrac{1}{2} - \dfrac{1}{2} = 0$.

The expression $S = 0$ is known as a singlet transition, and if the coupled angular momentum of the two leptons $L = l_e \mp l_v = 0$, the J^π values of decaying nucleus and residual nucleus would be the same. The equation $S = 1$ is known as a triplet transition, and if $L = l_e \pm l_v = 0$, the J^π of initial and final nucleus may change by one unit such that $\Delta J = \pm 1$ or no change of parity.

For allowed and super allowed β decay transitions, one has the selection rules as follow $S = 0$, $L = 0$, $\Delta J = 0$, and no parity change between initial and final states.

For $S = 1$, $L = 0$, $\Delta J = \pm 1$, and no parity change between initial and final states will give rise to allowed decays.

For $S = 0$, β decay transitions are called fermi interaction and for $S = 1$, β decay transitions are called Gamow-Teller interactions. The two types of transitions are observed to proceed with approximately the same speed. The β decay transitions 0^+ to 0^+ are fermi type of transitions whereas n (½) to p (½) decays will produce both $S = 0$ and $S = 1$ and give rise to mixtures of both fermi and Gamow-Teller transition.

It is customary to quote the values of $\log ft$ for β decay rather than the value of matrix elements. Relationship between ft values and (M_{if}) are only valid for allowed β transitions. The ft values for forbidden transitions are much higher by many orders of magnitudes. Therefore, these values are expressed as $\log ft$ values.

In summary, one finds the following features of β decay:

1. There is a great variation in $t_{\frac{1}{2}}$, varying form 10^{-8} sec to about 10^{15} years as compared to nuclear time scale of 10^{-22} sec.
2. Lifetimes depend upon the value of angular momentum (L) carried by the (e^+, υ) pair, which depends upon the J^π values of the parent and daughter nuclei.
3. Lifetimes also depend upon whether parity changes or not.
4. Lifetimes also depend upon the β decay energy.

5.   It also depends upon the effect of Coulomb field on the wave function of the emitted e^-. Hence, the lifetimes would depend upon the Z-value of the nucleus.

In beta decay, electron and neutrino could be emitted with angular momentum values $L > 0$, i.e., L = 1, 2, etc.

Plane waves of e^- and υ can be expanded in a series with terms for different values of L.

Table 5.4. Classification of beta transitions according to log ft values

Types of Transition	Log ft	Selection Rules	Parity change
Super-allowed	3.5	$\Delta J = 0, \pm 1$	(No)
Allowed	5.5 ± 1.3	$\Delta J = 0, \pm 1$	(No)
First Forbidden	7.3 ± 1.3	$\Delta J = 0, \pm 1$	(Yes)
Second Forbidden	12	$\Delta J = \pm 2$	(No)
Third Forbidden	16	$\Delta J = \pm 2$	(Yes)
Fourth Forbidden	21	$\Delta J = \pm 2$	(No)

The type of transitions are labeled allowed and various degrees of forbidden, their respective log ft values and change of angular momentum and change of parities are given in table 5.4. Decays with log ft values between 3 and 4 are definitely allowed transitions. Some decays with log ft values greater than 4 are also allowed transitions whereas decays with values of 5.5 to 21 are forbidden with various degrees of forbiddeness. These transitions have long lifetimes. The decay rate of β decay depends upon the angular momentum change or selection rules as well as on the overlap between the wave functions of initial and final states. The higher angular momentum waves have very small amplitudes inside the nuclear volume and will therefore have a longer lifetime.

As discussed earlier, some β decays take place in addition to ground state to an excited state of residual nucleus, which means two end point energies of electrons. In such cases, the β-spectrum is quite complex.

5.23. Parity Violation or Nonconservation in Beta Decay

Any nuclear decay or reaction must obey certain laws of conservation, such as conservation of energy, linear and angular momentum, and conservation of parity, which is known as left-right symmetry.

In 1956, T. D. Lee and C. N. Yang (5.15) pointed out that there did not exist any experimental proof of parity conservation in beta decay, which is brought about by weak nuclear interaction. They suggested several experiments to test this law.

C. S. Wu and collaborators (5.16) decided to test this by the study of β decay of Co^{60}, which decays predominantly (99%) to the 2.50 MeV excited state of Ni^{60} nucleus with a change of intrinsic spin $\Delta J = 1$, $\Delta \pi = 0$ as shown in figure 5.14. The excited state then emits two gamma rays in cascade to the ground state of Ni ^{60}.

A normal source of Co^{60} would emit β^- particle equally in all directions; however, if the nuclei of Co^{60} are polarized, that is, oriented in a preferred direction, the resultant emission of beta particles will not be isotropic.

Wu and collaborators used low temperatures and external magnetic field to align the Co^{60} nuclei (T = 0.01^0K, magnetic field B = 10 gauss). The low temperature was created by adiabatic demagnetization of a cerium-magnesium nitrate crystal. As a paramagnetic ion, cobalt has a very high magnetic moment and an extremely strong internal magnetic field at the nucleus, thereby aligning the cobalt nuclei so long as the low temperature can be maintained (about 7-8 minutes).

Two beta counters were located at 90^0 to record the β-particles as a function of time. The results are shown in figure 5.15.

The above result indicates that more β-particles were emitted in the direction antiparallel to the angular momentum vector than parallel to it. Reading of two gamma ray counters showed an anisotropy, which fell from its initial maximum value of 1.2 to 0.75, a difference of 0.45 as shown in the figure corresponding to a maximum polarization of P = 0.65 falling to zero within 7 minutes

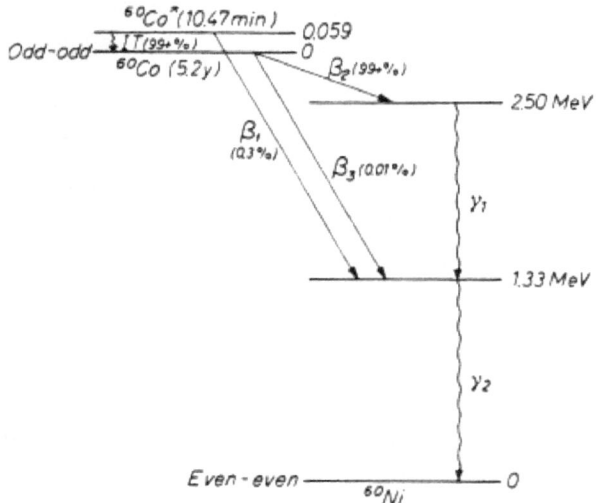

Decay scheme for ^{60}Co. Data on the decay are given in Table 50.2.

Figure 5.14. Decay scheme for Co^{60}

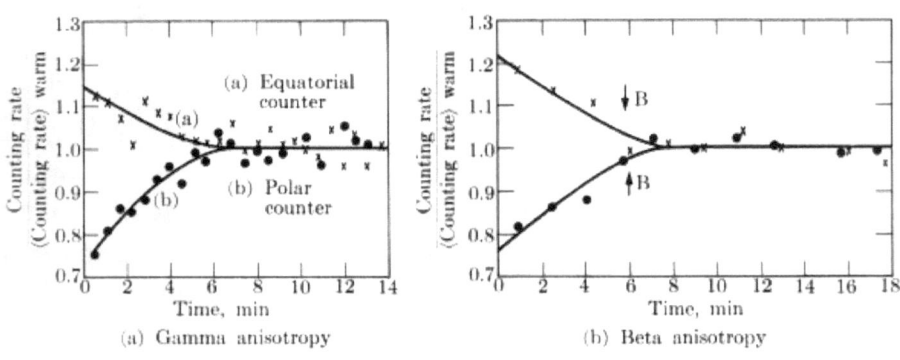

(a) Gamma anisotropy   (b) Beta anisotropy

Results of parity violation experiment. [From C. S. Wu *et al.*, *Phys. Rev.* **105**, 1413 (1957).]

The result shows that β emission takes place preferentially in the direction antiparallel to the nuclear spin orientation whereas parity conservation would require that β intensity antiparallel to the nuclear spin direction should be the same as along parallel-to-spin direction.

This shows violation of parity conservation. One believes that antineutrino is the culprit for parity violation.

5.24. Double β Decay

Sometimes a β-unstable nucleus is prevented from decay to its neighbor isobar due to energy consideration or change of spin and parity between the parent and daughter nuclei. One knows that change of $\Delta J = \pm 1$ with change of parity or $\Delta J = \pm 2$ and greater are strongly forbidden transitions. In such a case, if the β decay of the parent nucleus to its isobar is permitted energetically and requires change of angular momentum $\Delta J = 0$ or ± 1, it would be an allowed or super allowed transition. This decay will take place directly with the emission of two electrons or two positrons as the case may be accompanied by the emission of two neutrinos or antineutrinos. This decay is known as double β decay.

An example of double beta decay is decay of $_{52}\text{Te}^{128}$ to $_{54}\text{Xe}^{128}$ where decay to the neighbor isobar $_{53}\text{I}^{128}$ is energetically forbidden due to a negative Q- value of -1.26 MeV. However, double $\beta\beta$ decay $_{52}\text{Te}^{128}$ to $_{54}\text{Xe}^{128}$ is energetically possible with a Q-value of 0.87 MeV.

In most cases, the lifetimes of such decays range from 10^{15} to 10^{25} years. Therefore, such double beta decays are not observed experimentally except in rare cases.

5.25. Nucleon Emission Followed by β Decay

In general, β decay between neighbor isobars leading to either the ground or excited states of daughter nucleus have small Q- values. Thus, the excitation energies are below the threshold of nucleon emission. However, in cases where the excitation energy is greater than the nucleon threshold, the daughter nucleus will emit a nucleon.

In the fission of U^{235} with neutrons, fission fragments are left in excited states with energies above the threshold of neutron emission. In such case, delayed neutron emission has been observed, and this phenomenon is used in the design of nuclear reactors.

5.26. Electron Capture

Process of electron capture was first suggested by Yukawa and Sakata (5.19).

In this process, an atomic electron in one of the inner shells (K, L, etc.) finds itself inside the nuclear volume and is captured by a proton, which then changes to a neutron and emits a neutrino (v). This process is similar to the inverse process of β decay and competes with β^+ decay. The capture of electron is given as

$$p^+ + e^- = n + v \qquad (5.33)$$

Neutrino carries all the energy with negligible energy imparted to the recoiling heavy nucleus. Thus, the neutrino energy is discreetly defined by the decay. Monoenergetic neutrinos are emitted with kinetic energy given by

$$E_v = E_0 + m_e c^2 - E_B \qquad (5.34)$$

where E_0 is the available energy corresponding to mass difference between the parent and daughter nuclei. The E_B is the binding energy of the electron in the atomic shell. Theory of electron capture is very similar to the theory of β decay. Hence, it will not be discussed in this book.

5.26(b). Summary of β- decay

β-decay has played a very important role in nuclear physics. For example

(1) β-decay led to the discovery of neutrino, a charge less and mass less particle.
(2) The study of β-decay also led to a new type of nuclear force, the weak force about 10^{-13} times weaker than the strong nuclear force.
(3) β-decay also led to the discovery of non conservation of parity in weak interactions.

In 1961 Lederman and collaborators discovered two types of neutrinos, one accompanied by electrons in the β- decay and the other emitted in the muon decay. In the universe billions of stars producing energy from fusion of nuclei emits large number of neutrinos. Neutrinos are thus considered likely candidates for the missing dark matter in the universe.

Theory of β- decay is very complex. Probability of β- decay depends upon many factors, such as angular momentum carried by the pair electron and anti-neutrino and the change of parity between the parent and daughter nuclei.

β- decay also depends upon the overlap of wave functions of the parent and daughter nuclei

Some decays are super allowed with very short **half lives and** some are extremely forbidden with very long lives.

Scientists predicted that the particles mediating β - decay and weak interactions are $W^{\pm}$ and Z^0. These particles emitted in high energy p,p collisions were observed in 1983 at CERN.

5.27. Gamma Decay of Unstable Nuclei

Theory of electromagnetic radiation emitted by atoms was developed by Niels Bohr in 1913 by describing atoms consisting of a nucleus at the center and electrons revolving around the nucleus as a result of attractive electric force. Electrons were assumed to be located in different shells, rotating in fixed orbits. Electromagnetic radiation (light) was emitted when electrons jump between different orbits.

Similar concepts are involved in the emission of gamma rays by unstable nuclei. According to shell model of the nucleus, protons and neutrons are rotating in orbits in fixed shells as a result of average central potential created by nucleon-nucleon force.

According to shell model of the nuclear structure, when a nucleus absorbs energy by some process, the nucleons move from a lower shell to a higher shell and give rise to an excited state. When these nucleons return to their original orbits, excess energy is released in the form of gamma ray.

In medium and heavyweight nuclei, energy is absorbed and emitted in a collective motion of many or all nucleons instead of a single nucleon as is the case for a shell-model nucleus. The emission of electromagnetic radiation is caused by the changes in the energy of collective excitation.

In general, a highly excited state of a nucleus will de-excite by the emission of a nucleon whenever it is energetically possible. Below the dissociation energy of a nucleon, de-excitation of state takes place by electromagnetic interaction accompanied by a gamma ray emission.

5.28. The Objective of Gamma Ray Studies
Involve the following:

1. Comparison of experimental data on excited states energies and J^{π} with the theoretical prediction based on different models of nuclear structure.
2. This study allows one to determine the important level parameters such as its energy, process of de-excitation giving information about the electromagnetic transition probabilities between states.
3. Higher energy levels in nuclei can be excited by the absorption of nucleons to energies equal to the binding energy of nucleons

added with the incident energy. These states can also be excited from inelastic scattering of nucleons, which leaves the residual nucleus in an excited state. In most cases, an excited state emits many gamma rays in cascade.

5.29. Measurement of Gamma Ray Energies

For measurement of gamma ray energies, one uses many different types of detectors. The most popular detector used is a NaI(Tl) scintillator coupled with a photomultiplier. In this detector, gamma rays interact with the NaI scintillator and emits photoelectrons, which carry γ-ray energy and emits light in the scintillator, which is transmitted via a light guide onto a photocathode of the photomultiplier. This light emits photoelectron from the cathode, and after acceleration strikes, various metal dynodes located inside the photomultiplier. Each time the electrons strike the dynode, these eject more electrons and very quickly electrons multiply to a large number. This avalanche of electrons are collected at the anode, which has a potential difference between the anode and the cathode of about 2,000 volts. An electronic pulse is generated in a capacitor and fed to a pulse height analyzer. The current pulse whose height is proportional to the gamma ray energy is analyzed. This pulse height is calibrated with a source of known γ-ray energy.

The scintillator photomultiplier combination produces three pulses of different heights for each gamma ray energy. These pulses are known as photo, compton, and pair peaks.

5.30. Compton Effect

When a gamma ray interacts with an atomic electron of the NaI detector, the γ-ray is scattered inelastically, and it imparts the balance energy to the electron. The electron's energy depends upon its angle of scattering. Hence, a continuous range of energies are carried by the scattered electron. These electrons produce light in the scintillator proportional to the energies of scattered electrons. This light produces photoelectrons in the cathode and finally generates an electronic pulse proportional to the energies of compton electrons. This spectrum is flat.

5.31. Pair Production

In addition to these above-mentioned processes of interaction with the detector, there is another process of interaction. If the γ ray energy is greater than 1.02 MeV, this gamma ray interacts with the atoms of iodine (high Z element) and creates a pair of positron and electron known as pair production. These pair then annihilates producing two 0.51MeV γ-rays emitted in opposite directions. In general, these γ-rays escape from the scintillator. Thus, an electronic pulse proportional to E_γ- 1.02 MeV energy is generated at the anode.

When a nucleus is excited to high energy, it emits several γ rays of different energies. Hence, the spectrum produced by the multiplier is very complex and quite often different peaks are overlapping. This type of detector suffers from poor energy resolution but has high efficiency due to its size which can be varied at will.

5.32. Ge (Li) Solid State Detector

During 1960, a new type of solid state detector known as Ge (Li) detector was developed. The detector consists of a p-type germanium wafer in which lithium atoms are deposited to form an n-type region. Under reverse bias and slightly elevated temperature, Li atoms drift into p-type region, creating a large depletion layer. This detector is operated at low temperature to prevent lithium atoms migrating out of its lattice sites.

The detector comes in different sizes to produce greater efficiency. This detector has about 10-20 times better energy resolution than the NaI detector. The γ ray energies that overlap in NaI detectors are clearly resolved in this detector. This detector is commonly used in the measurement of γ ray energies.

Figure 5.15b shows a spectrum taken by this detector for a γ ray energy of 2.75 MeV. The spectrum shows a photo-peak, a compton broad peak, and a pair production peak. The energy resolution as seen in the figure for 2.75 MeV γ rays is 2 parts in 1,000.

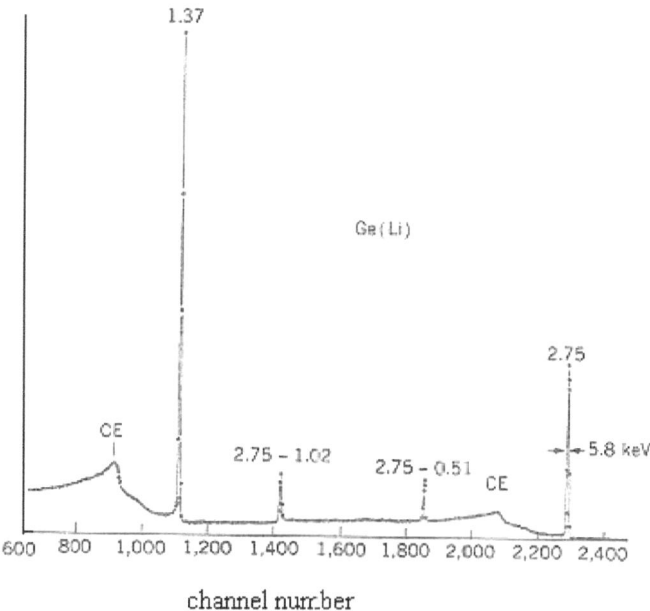

Figure 5.16. The figure above shows the spectrum taken by Ge (Li) detector of 2.75 MeV γ ray.

5.33. Bent Crystal Spectrometer

Another type of detector used for the measurement of γ ray energies is bent crystal spectrometer. This detector measures the wave length or frequency of γ rays. Since energy is related to the wavelength by the relation E = hc/λ, one can determine the energy if one knows the wavelength. The device is based upon x-ray Bragg diffraction method. In this detector, the source of γ rays is placed inside a bent quartz crystal. A detector is placed on the other side of the crystal behind a lead collimator. The Bragg-diffracted γ ray is counted by the detector, and the angle is determined.

By using the formula 2d sin Θ = nλ, one can determine the value of λ, and, hence, the γ ray energy.

This type of detector has very high energy resolution, and the accuracy of energy is 1 part in 10,000 for energy of gamma rays larger than 100 keV.

5.34. Selection Rules

If $J_i^{\pi_i} \rightarrow J_f^{\pi_f}$ are the total spin and parities of the initial and final states of an excited nucleus as shown below, a transition from the initial state to the final state would take place by the emission of a γ ray of energy equal to the difference of energies between the two states.

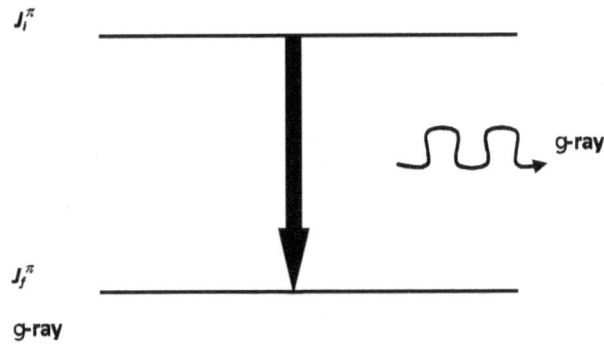

Excited Nucleus

The multipolarity of γ-ray transition is determined by the value of L that is transferred to the gamma radiation such that

$$\left|J_i - J_f\right| \le L \le J_i + J_f$$

The parity change is determined by the L value of the gamma radiation. For an electric transition (EL), $\pi_i\pi_f = (-1)^L$ and for a magnetic transition (ML), 2^L pole radiation is associated with the parity change as $\pi_i\pi_f = (-1)^{L+1}$.

Lifetime $(t_{\frac{1}{2}})$ of the transition depends upon the multipolarity of gamma rays and upon the nuclear model responsible for the excited states. There are no monopole transitions (L = 0). Therefore, a transition between $O^+ \rightarrow O^+$ is forbidden.

5.35. Lifetimes of Gamma Ray Transitions

When the excited energy of a nucleus is higher than the dissociation energy of a neutron or proton, the nucleus will always decay by particle emission. This decay is extremely fast with $t_{\frac{1}{2}} \sim 10^{-20}$ sec .

Below the dissociation energy, the excited states decay by electromagnetic radiation to either the ground state directly or via cascade γ-ray transmissions via intermediate states. The lifetimes $t_{\frac{1}{2}}$ of such decays are much longer than the lifetime of a nucleon emission.

Quantum mechanical states that are not completely stable do not have sharp values of the energy.

Uncertainty principle gives $\Delta E.\Delta t \leq \hbar$.

If $t_{\frac{1}{2}}$ is the lifetime of states, the energy width of the decaying state is

$$\Gamma = \frac{\hbar}{t_{\frac{1}{2}}} = \frac{6.58 \times 10^{-16}}{t_{\frac{1}{2}}} (eV)$$

One can get a value of $t_{\frac{1}{2}}$ from the measurement of Γ value or line width. A mean life of 10^{-17} sec would give $\Gamma = 65.8$ eV. Such widths have been measured for many gamma ray transitions all across the periodic table.

5.36. Theory of Gamma Ray Decay

A complete description of the emission and absorption of electromagnetic radiation by a nucleus requires quantum theory of radiation. However, useful information can be obtained from a classical discussion based upon the fact that protons and neutrons have charges. It is known that distribution of electric charge inside a nucleus produces electrical potential (V), which at large distances R in the Z direction can be expended as

$$V = 1/4\pi\varepsilon_\circ (1/R \int \rho \, dV + 1/R^2 \int \rho Za \, dV + 1/R^3 \int \rho(3Z^2 - V^2) dV$$

where ρ is the charge density and the integral is over the volume V of the nucleus containing the charge. In view of the powers of R in the denominator, this is a rapidly converging series. The first term is the potential due to the total charge. These distributions of charge produce electric moments, and ensuing electric currents produce magnetic moments. Second term gives rise to dipole moment, and the last term produces quadrupole moment. Magnetic moments of nuclei are complex and depend upon the angular moment and spin moment of the nucleus.

Theory of gamma ray emission is based upon the fact that electromagnetic radiation is produced by an oscillating electric and magnetic field propagated in space in the form of a transverse waves.

Protons or neutrons inside the nucleus carrying electric and magnetic charges are set into oscillation after being excited and propagating electric and magnetic fields and energy in space.

5.37. Classical Theory of Gamma Rradiation

We know from classical physics that a current flowing in an antenna generates an electromagnetic periodic transverse wave-producing electric and magnetic fields perpendicular to each other and to the direction of the propagation of the wave. Electric and magnetic fields vary as a function of distance r and as a function of time (t). The motion of these fields is described by Maxwell's theory, which states that in vacuum, each component of the electric field E and magnetic field B satisfies the equation

$$\nabla^2 \psi = \frac{1}{c^2} \frac{\partial^2 \psi}{\partial t^2}$$

The solution of this equation gives electric field as

$$E\,(r,\,t) = E\,(r)\,e^{-\omega t} + E^*(r)\,e^{\omega t}$$

and the magnetic field as

$$B\,(r,\,t) = B\,(r)\,e^{-\omega t} + B^*(r)\,e^{\omega t}$$

where ω is the angular frequency of the oscillator and total energy density(U) is given as

$$U = \tfrac{1}{2}\,\varepsilon_o\,E^{\,2} + \tfrac{1}{2}\,B^2/\mu.$$

Electric and magnetic fields are subdivided into electric and magnetic multipoles. General solutions of these equations involve spherical hankel functions and spherical harmonics $Y_{lm}(\theta,\varphi)$. This discussion is quite complex, and hence, we quote here the expression for the power radiated by the antenna.

Radiated power for electric multiple moment Q_{lm} oscillations with an angular frequency (w) is given as

$$P_E(l,m) = \frac{2(l+1)C}{\varepsilon_0\, l\left[(2l+1)!!\right]^2}\left(\frac{w}{c}\right)^{2l+2}\left|Q_{l,m}\right|^2 \; watts \qquad (1) \qquad\qquad (5.35)$$

and power radiated for magnetic moment M_{lm} is given as

$$P_M(l,m) = \frac{2(l+1)C\mu_0}{l\left[(2l+1)!!\right]^2}\left(\frac{w}{c}\right)^{2l+2}\left|M_{i,m}\right|^2 \; watts \qquad (2) \qquad\qquad (5.36)$$

where double factorial 7!! means $= 1\cdot3\cdot5\cdot7\ldots$ and 8!! means $= 2\cdot4\cdot6\cdot 8\ldots$

For electric dipole radiation designated as E1 with $l=1$ and $m=1$

$$P_E(l,m) = \frac{4w_0^4}{9\varepsilon_o\,c^3}\left|Q_{lm}\right|^2 \; watts \qquad\qquad (5.37)$$

Values of multipole moments $(Q_{l,m})$ for $(1,1), (1,0), (2,0)$ are given below.

$$Q_{11} = \sqrt{\frac{3}{8\pi}}(x-iy)\rho\,d\tau = \sqrt{\frac{3}{8\pi}}er$$

$$\left|Q_{11}\right|^2 = \left(\frac{3}{8\pi}\right)e^2r^2$$

$$Q_{2,0} = \sqrt{\frac{5}{16\pi}}\int(3z^2 - r^2)\rho\,d\tau$$

If r = radius of the nucleus and w_0 the angular frequency, the centripetal acceleration is $a_c = rw_0^2$. Substituting the value of l and Q_{11} and a_c in 5.37, one gets the value of radiated power as

$$P_E(1,1) = \frac{w_0^4 e^2 r^2}{6\pi \in_0 c^3} = \frac{e^2 a^2}{6\pi \in_0 c^3} \text{ watts}.$$ (5.38)

As the proton radiates power, its energy decreases and the radius decreases. Assuming that the charge moves in a harmonic oscillator potential, its total mechanical energy is given as

$$E = mw_0^2 r^2$$

$$E = pt \quad \text{where } p \text{ is power and } t = \text{time}.$$

Using $\quad E = \left(\frac{w_0^4 e^2 r^2}{6\pi \in_0 c^3} \right) t$

$$\frac{dE}{dt} = -P_E(1,1) = -\frac{w_0^4 e^2 r^2}{6\pi \in_0 c^3}$$

Substituting value of $E = mw_0^2 r^2$ in the above equation, one gets

$$\frac{dE}{dt} = -\frac{w_0^2 e^2 E}{6\pi \in_0 mc^3}$$ (5.39)

$E = E_0 e^{-\lambda t}$ (Energy decays exponentially with time)

$$\frac{E}{t} = \lambda(E_L) = \frac{w_0^2 e^2}{6\pi \in_0 mc^3} \sec^{-1}$$ (5.40)

5.38. Quantum Theory of γ Radiation

Transition to quantum mechanical treatment requires the fact that in classical mechanics, energy radiated is continuous whereas in quantum mechanics energy is quantized and is given as $E = \hbar\omega$ where ω is the angular frequency of the harmonic oscillatcr. Secondly, the source of radiation is due to the motion of protons and neutrons, which obey laws of quantum mechanics.

According to quantum theory, the emission of electromagnetic radiation takes place from an initial state to a final state of the nucleus. These states have well-defined energies, total angular momentum J and parities. When a γ-ray is emitted, it carries an angular momentum L, which determines the change of parities between states. The value of momentum L of γ-ray is determined by the law of conservation of angular momentum and parity change of the system.

Transition probability of γ ray from one state to another state of nuclei can be described as

$$\lambda = \frac{2\pi}{\hbar^2}|M_{if}|^2 \tag{5.41}$$

where M_{if} is the matrix element describing the overlap of final and initial state-wave functions, which is given as

$$M_{if} = \int \psi_{fn}^* H \psi_{in} d\tau \tag{5.42}$$

where H is the energy operator related to electromagnetic interaction.

The term $L = mw_0^2 r^2$ is the angular momentum carried away by the radiation field as the energy decays to zero. Where $w=$ frequency of wave, the radiated energy is quantized, and angular momentum value is

$$L = \hbar\sqrt{L(L+1)}$$

$$\lambda_E(l=1) = \frac{2(l+1)}{l\left[(2l+1)!!\right]^2}\left(\frac{3}{l+3}\right)^2 \frac{e^2 R^{2l}}{4\lambda\epsilon_0\,\hbar}\left(\frac{w}{c}\right)^{2l+1} \text{sec}^{-1} \tag{5.43}$$

$$\lambda_M(l=1) = \frac{20(l+1)}{l\left[(2l+1)!!\right]^2}\left(\frac{3}{l+3}\right)^2 \frac{e^2\mu_0 cw}{4\pi\hbar}\left(\frac{\hbar}{mcR}\right)^2\left(\frac{wR}{c}\right)^{2l} \text{sec}^{-1} \tag{5.44}$$

Substituting the values of gamma ray energy in MeV and R in fermi, one gets

$$\lambda_E (l=1) = \frac{4.4(l+1)10^{21}}{l\left[(2l+1)!!\right]^2}\left(\frac{3}{l+3}\right)^2\left(\frac{E_r}{197}\right)^{2l+1} R^{2l} \sec^{-1}. \tag{5.45}$$

A similar calculation for a magnetic transition is given as

$$\lambda_M (l=1) = \frac{1.9(l+1)10^{21}}{l\left[(2l+1)!!\right]^2}\left(\frac{3}{l+3}\right)^2\left(\frac{E_r}{197}\right)^{2l+1} R^{2l-2} \sec^{-1} \tag{5.46}$$

Ratio of electric to magnetic decay probabilities are given as

$$\frac{\lambda(E1)}{\lambda(M1)} = \frac{1}{10}\left(\frac{M_p Rc}{\hbar}\right)^2, \text{ where } M_p = \text{ proton mass}. \tag{5.47}$$

Selection rules for the emission of gamma rays of electric and magnetic multipoles are given below in table 5.2.

Table 5.2

Type	symbol	l	parity change
Electric Dipole	E1	1	yes
Magnetic dipole	M1	1	no
Electric Quadrupole	E2	2	no
Magnetic Quadrupole	M2	2	yes
Electric Octupole	E3	3	Yes
Magnetic Octupole	M3	3	no

In general, transition probability decreases rapidly as L value increases. For all practical purposes, we can ignore transitions with $L > 3$.

In general, more than one type of radiation is possible between two states. For example, if angular momentum and parities of the initial and final states are $J_i = 1^+$ and $J_f = 2^-$, the possible L-values of γ rays are $L = 1, 2$, and 3. The parity of initial state is (+), and the final state is (-). This means that there is a change of parity. Hence, for the **possible** values of L and parity change, the multipolarities of γ rays would be E1, M2, or E3, since the transitions probability decreases rapidly with increasing L-value; hence, for practical purposes, one can neglect transitions with $L > 1$, or in this case, the most probable transition would be E1 with a small mixture of M2.

Values of $\log \lambda$ as a function of gamma ray energy for both electric and magnetic multipole transitions based upon single particle transitions have been calculated by Victor Weisskopf and S. Moskowski. These are plotted by Edward Condon and Hugh Odishaw (5.17) and are shown in figures 5.16 and 5.17.

On the left of the figures are given $\log \lambda$ for different L-values as a function of E_λ; the right of these figures gives level width $\Gamma = h/\tau_{1/2}$ in eV.

Estimation of radiation widths (Γ_γ) by Weisskopff (5.18) for γ-ray transitions with L =1 and 2 are given as follows:

$$\Gamma_\gamma (E1) = 6.8 \times 10^{-2}\, A^{2/3}\, E^{3}\ \text{ev}$$
$$\Gamma_\gamma (E2) = 4.9 \times 10^{-8}\, A^{4/3}\, E^{5}\ \text{ev}$$

The corresponding widths for magnetic transitions are as follows:

$$\Gamma_\gamma (M1) = 2.1 \times 10^{-2}\, E^{3}\ \text{ev.}$$
$$\Gamma_\gamma (M2) = 1.5 \times 10^{-8\ A\,2/3}\, E^{5}\ \text{ev}$$

The decay probability varies by many orders of magnitude as the gamma ray angular momentum (L) changes by one unit. This ratio decreases as the gamma ray energy increases.

Weisskopf (5.18) has used shell model wave functions to estimate transition probabilities. The experimental data shows that γ-ray emission from excited states of nuclei cannot very well be described in terms of pure single-particle excitation nucleus so that more than one single unit of charge takes part in the transitions providing greater probability of γ-ray transitions. However, transitions for nuclei in the region of large quadrupole moments E2 transitions are observed to be more than one hundred times faster than the single particle estimates. These nuclei-display collective excitations, i.e, rotation or vibration of nuclei.

There is no experimental method to determine directly between different types of multipole radiation by observation of gamma rays emitted from nonoriented nuclei.

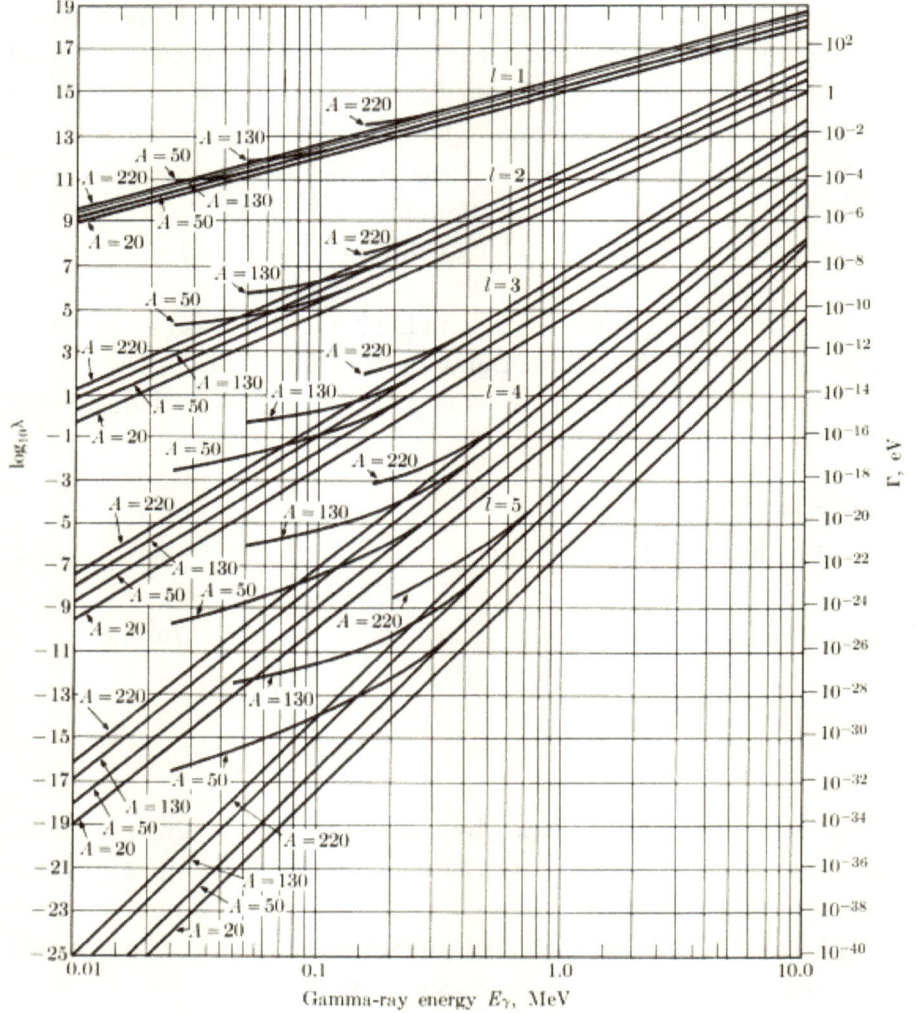

Figure (5.17) Probability for electric multipole transition based on Weisskopf's and Moszkowski's single-proton estimates. The curved lines indicate the total transition probabilities with internal conversion included. (From E. U. Condon and H. Odishaw, *Handbook of Physics*, New York: McGraw-Hill Book Company, 1958, pp. 9–109. Reproduced by permission.)

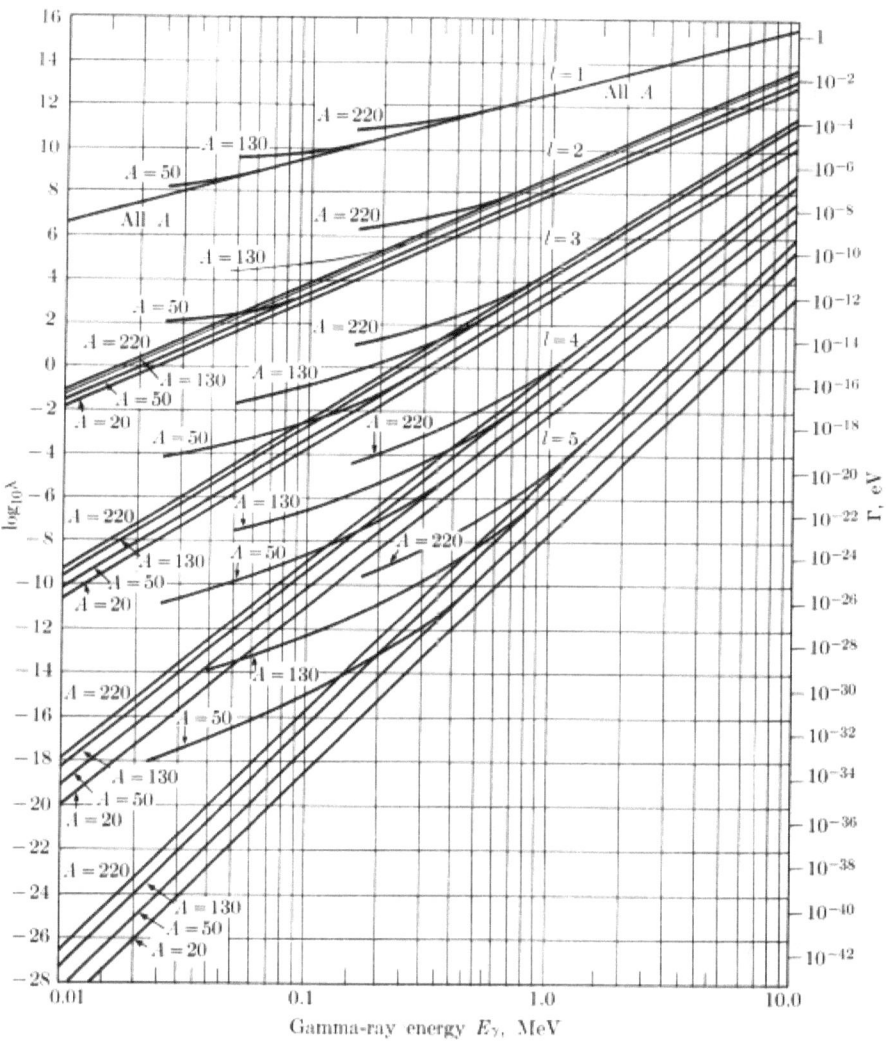

Figure 5.16. Probability for magnetic multipole transitions based on Moszkowski's single-proton estimates. Weisskopf's formula (Eq. 9–35) gives only slightly different transition rates. The curved lines indicate the total transition probabilities with internal conversion included. (From E. U. Condon and H. Odishaw, *op. cit.*, pp. 9–110. Reproduced by permission.)

However, if the nuclei are oriented so that the direction m_j of the angular momentum is known, the angular distribution of gamma rays is no longer spherically symmetric, and the shape of this distribution reveals the multipolarity L.

5.39. Angular Distribution of Gamma Radiation

For both the electric and magnetic transitions, gamma ray angular distribution as a function of the angle Θ is given as

For example, if $l = 1$ and $m_l = 0$, one has the γ-ray intensity as in the following:

$$S_{10} = \frac{3}{8\pi}\sin^2\theta \qquad\qquad (4.48)$$

And for $l = 1$, $m_l = 1$ one has the intensity distribution as in the following:

$$S_{11} = \frac{3}{16\pi}(1+\cos^2\theta). \qquad\qquad (4.49)$$

Angular distribution for $l > 1$ is anisotropic. Sometimes if the gamma decay takes place as a cascade of two gamma rays, one can measure the angular distribution of one gamma ray with respect to the other gamma ray. This angular distribution will depend upon the angular momentum carried by the gamma ray and the J^π of the initial and final states. Such measurements of angular correlation of gamma rays can provide valuable information about the spin and parities of decaying state.

5.40. Isomeric State

If the total angular momentum of the initial and final states are such that only higher angular momentum $L > 2$ transitions are permitted, such transitions have very long life about 0.1 second or **longer. Such** decaying states are known as isomeric states, and such a state decays mainly by electron conversion.

5.41. Internal Conversion

A nucleus in a bound excited state can also de-excite itself by a process called internal conversion. In this process, the available energy of the excited state instead of the emission of gamma ray is given to one of the orbital electrons of the atom, and that electron is ejected from the atom with the available energy. The electron could be from one of the atomic shells K, L, and M.

In general, a nucleus is left in the excited state by the emission of a β ray, such internal conversion electrons are emitted along with the β electron. Since the β electron have wide range of energies whereas the conversion electrons have well-defined unique energies associated with the orbital electron from different inner shells, the measured electron spectra shows discreet energy lines superimposed on a broad distribution of electron energies due to beta decay.

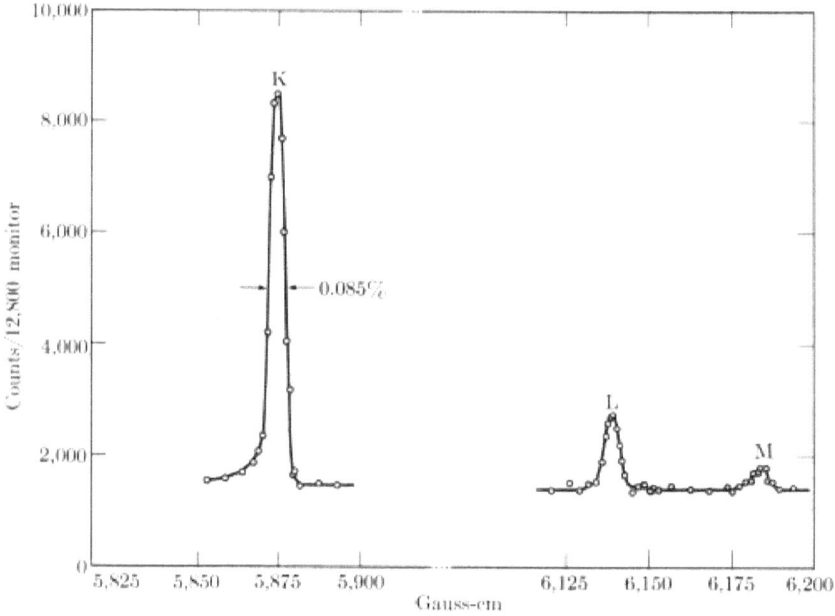

Figure 5.18. Internal lines from the 1.4158-MeV zero-to-zero (E0) transition in Po^{214}. [From D. E. Alburger and A. Hedtran, *Arkiv Fysik* 7, 424 (1953-54).]

conversion lines due to emission of K, L, and M electro**ns.** The gamma ray transition 0^+ to 0^+ is completely forbidden; therefore, the excited state decays by electron conversion. A typical spectra of conversion electrons due to Alburger et al. (5.19) from 1.4158 MeV transition in Po^{214} is shown in figure 5.18.

For all electromagnetic transitions, γ-ray emission competes with internal conversion whereas for 0^+ to 0^+ transition, internal conversion is the sole mechanism of decay. Probability of two competing process is given by the ratio $\alpha = \lambda_\gamma / \lambda_c$ where α is known as conversion coefficient.

The values of α have been determined for many cases, and it shows large variation in its value as a function of γ-ray energies.

In figure 5.18, one observes three
The kinetic energy of the electrons is given as

$$T = E_\gamma - E_x$$

where E_x is the binding energies of the K, L, and M shell.

5.42. Summary of Gamma Ray Transition Probabilities

Gamma rays are emitted from excited states of nuclei either to the ground state directly or through cascades to lower excited states and then to the ground state.

1. Energies of γ rays depend upon the energies of excited states above the ground state of a nucleus. These energies, in turn, depend upon the structure of nuclei in different mass region.
2. According to shell model, the excited states are due to a single nucleon making a transition from a lower shell to higher shells. Such states are observed in light mass nuclei or in closed-shell nuclei.
3. The excited states in medium and heavy nuclei are observed to have a regular pattern produced by the collective motion of many nucleons.
4. The γ-ray transition probabilities or half-life depends strongly upon the γ-ray energies varying by a factor of 100 for a change in energy from 1 MeV to 2 MeV.
5. The γ-ray transition probabilities depend strongly upon the angular momentum (L) carried by the gamma rays.

6. According to the shell model estimates of Weisskopf plotted in figure 5.16 for electric transition, the decay probabilities can vary by a factor of 10^5 for a change of one unit of change in L value at low energy. This factor becomes smaller for higher energy γ- rays.

7. Similar behavior of values of transition probability are seen for magnetic mutipole transitions plotted in figure 5.17.

8. For same value of L, the electric transitions have higher probabilities by a factor varying as A^2.

9. For highly deformed nuclei, the probabilities for E2 transitions are about 100 times greater than for shell model nuclei.

10. Similar effect is observed for medium weight nuclei where excited states are produced by collective vibration.

CHAPTER 6

NUCLEAR REACTIONS

6.1. Nuclear Reactions and Transmutation

All matter is made up of atoms. The nucleus, the central part of the atom, contains protons and neutrons. These nucleons are held together inside the nucleus by the strong attractive and short-range nuclear force. Protons have positive charges, which produces repulsive and long range Coulomb force. These two forces compete, but since the nuclear force is about 100 times stronger than the Coulomb force, this effect becomes appreciable only when large number of protons are involved. This effect causes nuclei to contain more neutrons than protons as the nucleus becomes heavier. When free nucleons such as neutrons and protons come close to each other within the range of nuclear force, these particles interact, causing a nuclear reaction.

Study of nuclear reaction is carried out by bombarding a target nucleus with an incident particle having kinetic energy.

6.2. Particle Accelerators

For this purpose, one needs to accelerate particles in an accelerator. Historically, before the invention of particle accelerator, nuclear reactions were induced by alpha particles from radioactive sources such as polonium nucleus. In 1919, Rutherford performed such a nuclear reaction by bombarding a target of N^{14} nucleus with α- particles and observed a transmutation of this nucleus with the formation of O^{17} nucleus and the emission of protons. The reaction is given as

$$_7N^{14} + _2He^4 = _8O^{17} + _1H^1.$$

Radioactive sources produce particles of several energies simultaneously, and therefore study of nuclear reactions us:ng these sources was not vey useful. For a better understanding of reaction mechanism, it was considered essential to produce particles with precise energy.

With this aim, Cockroft and Walton (6.1) developed the first direct current accelerator by generating 1 million volt across a terminal. Positively charged ions of hydrogen protons were accelerated inside a vacuum tube. This generated a maximum energy of protons of 1 MeV. These scientists bombarded a Li^7 target with energetic protons causing a nuclear reaction with the production of α particles.

With the development of high energy accelerators, scientists were able to accelerate different nucleons p, d, and α particles to high energies. The nature of nuclear reactions therefore changed considerably depending upon the particle energies and their type. Based upon the nature and type of nuclear reaction, nuclear physicists deve_oped many concepts and theories to explain reaction mechanism.

The basic components of a particle accelerator are the following:

a.   An ion source—Atoms of gases are ionized to produce positively charged particles.
b.   Electric and magnetic fields—This is provided by a magnet or an electric power.

Electric fields are produced by separating positive and negative charges in different locations like a battery or rectifying a radio frequency AC voltage. When a charged particle passes through such an electric field, it is accelerated to higher speeds and higher kinetic energies.

Magnets are used in general to focus the charged particle.

6.3.   Van de Graaff Accelerator

The next important developmen: in the acceleration of particles was made by R. J. van de Graaff at Massachusetts Institute of Technology (6.2). In this accelerator, positive charge produced at a corona point is carried on an insulated conveyor belt and then deposited on an insulated metallic dome. Enough charge is thus accumulated on the dome to produce a voltage difference of about 6 million volts between a positive terminal and the ground. This voltage is distributed uniformly throughout the length of the accelerator tube. Charged ions of protons or deuterons produced inside a gas bottle by striking an arc are accelerated inside a vacuum tube up to a maximum energy of 6 MeV. In order to avoid electric discharge at high voltage in air, the whole assembly is enclosed inside a tank filled with a high dielectric constant gas SF_6 at about twenty atmospheric pressures. One great property of this type of accelerator is that the energy of particle can be varied in small increments. Essentially, monoenergetic particles are produced. One drawback of this accelerator is that it provides low current in the range of few mA. During 1940-60, this type of accelerator was built at many universities and at national laboratories. Nuclear reactions induced by protons, deuterons, and α particles at many energies were investigated providing a wealth of information about reaction mechanism and properties of excited states of nuclei, etc.

With the success of this accelerator, scientists aimed to produce higher energies of particles.

6.4. Two-stage Tandem Accelerator

In this accelerator, two Van de Graaff accelerators are joined in tandem as shown in figure 6.2. In the first stage of this accelerator, positive ions are produced in an ionization chamber and are accelerated to some energy. These ions are then converted into negative ions by adding two electrons. As negative ions, these are accelerated in the first van de Graaff accelerator to a maximum energy of 6-10 MeV. These energetic negative ions emerging from the first stage pass through a gas stripper and are stripped of their electrons, and these become positive ions. The positive ions are then accelerated by the second Van de Graaff accelerator to an energy of 6-10 MeV. Thus, the final energy of the positive ions is about 12-20 MeV. The energetic particles are then deflected by a magnet to the desired location of the target to be bombarded. This type of accelerator provides particles in small increment of monoenergetic energies. These accelerators were expensive to build, and therefore only few universities and national laboratories were able to build these accelerators. At present, most of the studies in nuclear reactions are carried out by these types of accelerators.

6.5. Dynamitron Accelerator

A direct current accelerator built by Radiation Dynamics provides energy up to a maximum of 4 MeV but much higher beam currents in the range of few amps for protons. The accelerator uses RF oscillators and rectifiers to provide direct current voltages. The energies of particles can be changed in small increments. This type of accelerator was installed in 1970 under the direction of the author at the University of Albany, New York. It was mainly used for the study of application of nuclear physics in medical, industrial, and environment fields. This accelerator is also used for implantation of heavy ions in metals.

6.6. Cyclotrons

Around 1931, another type of particle accelerator was developed by scientist Ernest Lawrence at University of Berkeley (6.3). This accelerator consisted of two evacuated chambers in the shape of D joined together with a gap in between the chambers. The chambers were placed in between a homogenous magnet producing a magnetic field. An RF voltage was applied between the gaps. The positive ion source was placed at the center of the chambers. Ions emitted from the source were accelerated and were subjected to the magnetic field of the magnet. One knows that if a charged particle with an electric charge q is moving with a speed v perpendicular to the magnetic field B, it will experience a force, which will rotate the charged particle in a circular path of radius r given as

$$B q v = m v^2 / r$$

When the charged particle moving in a circular path of radius r arrives between the gap, it is subjected to the RF voltage. This produces an acceleration of the particle increasing its speed. This makes the particle to move in a circle of larger radius. Each time the particle arrives between the gap, it receives the acceleration. The process goes on until the particle energy is such that it moves in a circle of maximum radius. In this way, the charged particle acquires the energy of about 6 MeV. At this time, the particle is deflected by a magnet, and it is ejected out of the chamber to the location of the target to be bombarded.

It was later realized that this accelerator could only produce a maximum energy of 6 MeV. It was discovered that as the particle's energy was increasing, its mass was also increasing based upon the formula $E = mc^2$. As a result, the particle was slowing down, and it was arriving late between the gap. The particle speed was out of phase with the *RF* frequency.

6.7. Synchrocyclotron

To overcome this problem, scientists developed a design where *RF* frequency was modulated to compensate the late arrival of the particle. Such accelerators are known as synchrocyclotrons.

These accelerators were built at several universities. The Nevis synchrocyclotron at Columbia University used frequency modulation from 28 Mc at the time of injection of protons at zero orbit to 18 Mc when the 385 MeV proton beam reached 73.5 in maximum beam radius. The *RF* was operated at 60 cps. Since the energy of 385 MeV is greater than the rest mass energy of π-meson, when the accelerator bombarded a target with 385 MeV protons, it produced a copious supply of π- mesons. This opened a new field of research using π- mesons to study their interaction with nuclei. The accelerator provides pulsed beams of protons. The accelerator was also used to provide an intense source of pulsed neutrons. When the 385 MeV protons strike a lead target, it produces copious supply of fast neutrons, which are moderated to produce a wide range of energies of neutrons. Extensive high resolution measurements of neutron cross section at a wide range of energies were carried out by Rainwater and his colleagues using this accelerator.

6.8. Synchrotrons

Another type of accelerator was the synchrotron in which both the *RF* frequency and the magnetic field were varied to overcome the problem of increased mass of the accelerated particle. Initially, these types of accelerators produced energies in the range of 50-100 MeV, and these were installed at various universities and at national laboratories. A design known as isochronous cyclotron were built at Indiana University and at Michigan State University. These machines could accelerate protons to energies of about 100 MeV and heavier ions to about 25 MeV per nucleon.

With the advancement in the design of superconducting magnets with strong focusing properties, it became possible to reach much higher energies. The cyclotron used a vacuum chamber in the shape of a ring. The magnets were placed around the ring, and R F voltage was applied to accelerate the particles. This technique was successful in producing energies of protons in the range of billions of electron volts.

Further, scientists developed techniques to accelerate positively and negatively charged particles in opposite directions to collide, which produced greater energies by annihilating these particles in collision. These accelerators are known as colliders. Such largest accelerators exist at Fermi National Laboratory in Batavia, Illinois, and one at nuclear research center in Geneva, Switzerland. These accelerators are providing energies of protons in the trillion electron volts (TeV) range.

These accelerators provide opportunities to thousands of scientists from all over the world to conduct high energy experiments. Over the years, these accelerators have yielded a wealth of new information and knowledge about the interaction of elementary particles and quarks.

Since most of the past and present research in nuclear physics is carried out by accelerators providing energy below 100 MeV, I will not discuss the details of the physics involved in these very high energy particle accelerators.

6.9. Linear Accelerators

This type of accelerator consists of a series of metal tubes known as drift tubes with a small gap in between them. These drift tubes are connected to high powered *RF* voltage klystrons with frequencies in the range of 1,000-3,000 MHz. The acceleration of charged particle occurs at each gap. The length of each drift tube is increased as the energy and the velocity of particles increases.

In general, there are many such accelerators in use, but most of them accelerate electrons. The longest one-mile long accelerator is located at Stanford University. This accelerator can produce electron energy in the range of few GeV. This accelerator has yielded a wealth of information about the sizes of nuclei and has provided evidence that quarks do exist in groups.

One other use of such accelerator is that when high energy electrons strike a high z element such as lead or bismuth, these produce bremsstrahlung γ rays. Such γ rays produce a strong source of neutrons of different energies. Study of high resolution neutron interaction with many nuclei has been measured using this type of accelerator, and neutron time of flight techniques at Oak Ridge National Laboratory and at Renssealer Polytechnique Institute.

Nuclear reaction is a process in which a change in the composition and/or energy of a target nucleus is brought about through bombardment with a single nuclear particle or a composite of particles or an electromagnetic radiation.

The objectives of nuclear reaction studies are twofold.

Nuclear reactions constitute one of the most powerful tools of nuclear spectroscopy, i.e., it provides information about the properties of levels of nuclei such as their energies and J^{π} values, their modes of decay, and decay probabilities, etc. The information about these quantities lead to an understanding of nuclear structure based upon different nuclear models, which had been developed over the years and are discussed in chapter 4. The measurement of cross sections (probabilities) for nuclear reactions can lead to an understanding of the nature of interaction. Various theories of different types of nuclear reactions have been developed over the years by many scientists, and the experimental data can be compared with the prediction of such theories about the nuclear reaction mechanism.

When a target nucleus interacts with a particle of sufficient energy, there are many ways in which the reaction can proceed.

An example of a reaction of 6.0 MeV deuterons with K^{39} target is shown in figure 6.1.

Total energy released in the reaction is equal to sum of rest mass energies of K^{39}, H^{2}, and the kinetic energy of the incident deuteron minus the rest mass energy of the residual nucleus plus the particle or particles emitted in the reaction.

This energy is known as Q-value of the reaction.

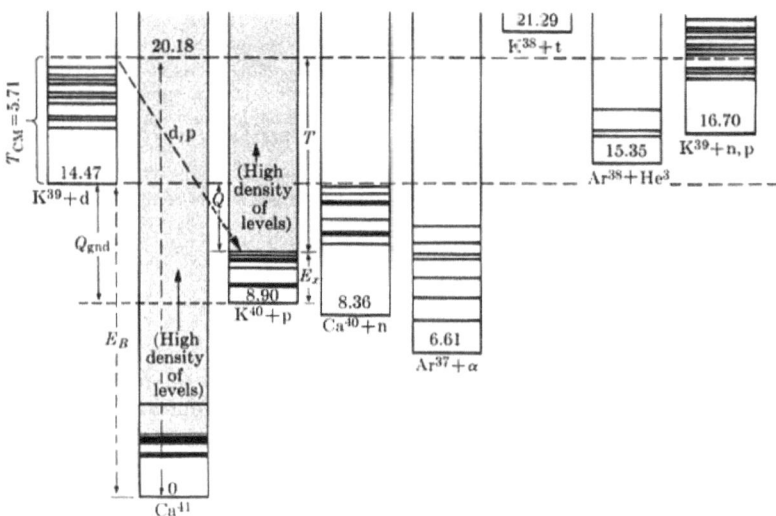

Figure 6.1   Energy diagram of nucleus plus particle systems with 20 protons and 21 neutrons.

As figure 6.1 shows that all reactions such as dd, dp, and dn, etc., that are energetically possible are likely to occur, for instance, those mentioned in the example shown above occur with different probabilities expressed as cross sections. When a residual nucleus is formed in the ground state, the available energy $Q_{gnd} + T_{CM}$ is given off as kinetic energy of the reaction products where T_{cm} is the kinetic energy of the deuteron in the center of mass.

In general, reactions involving emission of charged particles are subject to large reduction in cross sections due to Coulomb barrier, which the charged particle has to overcome to be emitted from the nucleus.

Reactions involving transfer of high angular momentum l are also inhibited, but there are no angular momentum or parity selection rules that completely prohibit certain reactions.

Tools of nuclear reaction studies can be classified into various components.

1.   Accelerators and reactors as sources of energetic neutrons, protons, deuterons, tritons, and alpha particles, etc.

2.   Detectors for detecting and measuring energies of different
     particles and gamma rays, etc.

3.   Appropriate equipment for making thin samples of nuclei used
     as target.

Energy region of greatest interest for nuclear reactions is 0-20MeV.

Particles that are used as projectiles are p, n, d, He^3, He^4, and gamma
rays. Neutrons having no charge can participate in nuclear reactions
essentially with zero energy. Charged particles mentioned above are
given energy by accelerating them in accelerators whose characteristic
has been discussed above.

A typical layout of a two-stage-tandem accelerator due to Van de
Graaff (6.4) is shown in figure 6.2.

In view of the fact that high voltage machine had better precision of
energy, which could be increased in small interval, these machines were
mostly used in measurements, which are discussed herein.

Height of Coulomb barrier for charged protons is given as

$$V_B = k\frac{Ze^2}{r} = 1.1ZA^{-\frac{1}{3}}\,(Mev),$$

where A is the atomic mass, and Z is the nucleon charge.

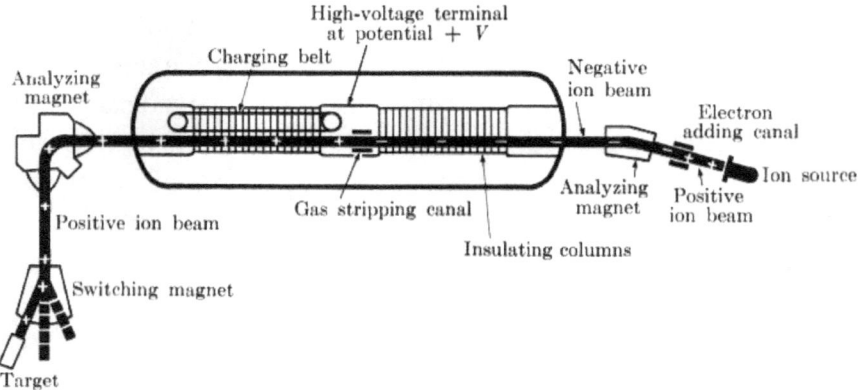

Figure 6.2.   A two-stage Tandem accelerator. [From R. J. Van de Graaff, *Nucl. Instr.*
Methods **8**, 195 (1960).]

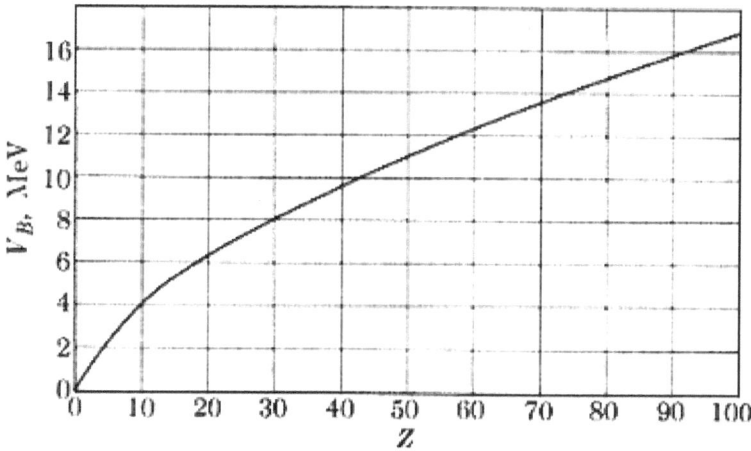

Figure (6.3) Height of the Coulomb barrier for proton, deuterons, and tritons as a function of Z

Plots of Coulomb barrier for protons, deuterons, and tritons—all having the same charge as a function of atomic mass A are shown in the figure 6.3. Barrier height varies with the charge of the incident particle as well as the Z of the target nucleus. For example, the Coulomb height for protons for a nucleus with $Z = 5$ is about 2.0 MeV and for $Z = 100$, it is about 17 MeV. These are the minimum energies of the protons needed to cause a nuclear reaction.

6.11. Nuclear Reaction and Law of Conservation

In any nuclear reaction, certain properties of the system remain constant. That is, these are conserved. Quantities that are conserved in a nuclear reaction are as follows: i = initial, f = final nuclei.

1. Total Energy $E_i = E_f$.
2. Linear Momentum $P_i = P_f$.
3. Angular Momentum $L_i = L_f$.
4. Total Charge $Q_i = Q_f$.
5. Parity = $(-1)^l$ is conserved in strong nuclear reactions, but not in weak interactions.

6. Statistics such as Bose-Einstien and Fermi-Dirac apply to the system of particles. Particles with spins $S = 1$ obey Bose-Einstein statistics and particles with $S = 1/2$ obey Fermi-Dirac statistics.
7. Baryon-number. Certain particles are known as baryons, and in a nuclear reaction their number is conserved.
8. Lepton number. Similarly certain light particles are known as leptons, and their number in any nuclear reaction is also conserved.
9. Isobaric-spin is conserved.

6.12. Broad Classification of Nuclear Reactions

Main types of nuclear reactions that have been studied in detail are as follows:

1. Resonance reactions or compound nucleus formation
2. Direct reactions—These have sub classifications
    a.) knock out reactions
    b.) stripping reactions
    c.) Pick up reactions
3. Other reactions, which do not fall in the above categories such as fission, fusion, Coulomb excitation, and photo nuclear reactions

The general basic theories of nuclear reactions are developed by many scientists such as the following:

1. Resonance Reaction given by Breit-Wigner formula (6.5) (based upon the compound nucleus formation)
2. R-matrix-formalism, formal theory of nuclear reactions
3. Direct reaction theories. Separate treatment are given for
    a. knock out reactions
    b. stripping reactions
    c. pick up reactions
4. Statistical theory of reaction
5. Photo-disintegration theory
6. Theories of nuclear fission and fusion

6.13. Experimental Observations Made in the Study of Nuclear Reactions are

1. Excitation functions ($(\sigma_T$ *Vs* $E)$. Cross section is measured as a function of incident-particle energy.
2. Resonances parameters $(E_R, \Gamma, \Gamma_n, \Gamma_r)$. E_R are energies at which resonance occurs, and Γ are the probabilities for each type of reaction such as elastic scattering, capture of particles, and for induced nuclear reaction.
3. Average cross sections (σ) over energy interval ΔE.
4. Strength functions of cross section averaged over an energy interval.
5. Angular distribution of particles emitted in the reaction. These distributions depend upon l-values of emitted particles.
6. Polarizations measurements of aligned nuclei.
7. Statistical effects based upon many body problems.
8. Nonstatistical effects related to nuclear structure.

6.14. Particle Detectors

For the study of nuclear reactions, it is necessary to detect the particles emitted in the nuclear reaction and to measure their energies accurately. Different types of detectors are used for different particles such as protons, deuterons, α particles, electrons, positrons, and γ rays. Commonly used detectors for charged particles and gamma rays are described in chapter 5 on radioactive decays.

6.15. Partial Wave Analysis of Reaction Cross Section

Consider a plane wave representing a particle beam incident on a nucleus located at the origin, and suppose the particles are neutrons with $l > 0$. The expansion in partial waves can be written in the following asymptotic form for large values of r as

$$\exp\left(ikr\cos\theta\right) \to \frac{\pi^{\frac{1}{2}}}{kr}\sum_{l=0}^{\infty}\sqrt{(2l+1)}\,i^{l+1}\left\{\exp\left[-i\left(kr-\frac{l\pi}{2}\right)\right]-\exp\left[i\left(kr-\frac{l\pi}{2}\right)\right]\right\}Y_{l,0}(\theta)$$

$$(6.1)$$

This is an undisturbed wave without the nuclear interaction being applied. When one turns on the nuclear interaction, the outgoing wave can change in phase and amplitude, but the incoming wave does not change. Let the outgoing part with angular momentum l change its amplitude by a factor η_l. This could be a complex number, but it is less than 1. The wave of scattered particle at large distance is given as follows:

$$\psi_{sc} = \psi_{total} - \exp(ikr\cos\theta)$$

$$\psi_{sc} = \frac{\pi^{\frac{1}{2}}}{kr} \sum_{l=0}^{\infty} \sqrt{(2l+1)}\, i^{l+1} (1-n_l) \exp\left[i\left(kr - \frac{l\pi}{2} \right) \right] Y_{l,0}(\theta,\phi) \qquad (6.2)$$

Since all particles in the scattered wave have same radial velocity (v), the flux, number of particles per unit area per second in this wave is $|\psi_{sc}|^2\, v$, where v is the velocity of particle equal to the flux of incident particles. Integrating spherical harmonic function over all angles gives the following:

$$\int_0^{2\pi} Y_{l,o}(\theta,\phi) = 1 \text{ since Ys are normalized function.}$$

Total number of particles per second in the scattered wave moving out through a shell of radius r in angular momentum state l is as follows:

$$N_{sc}(l) = \frac{\pi}{(kr)^2}(2l+1)|1-\eta_l|^2 \int_0^{2\pi} \left[Y_{l,0}(\theta) \right]^2 r^2 \sin\theta\, d\theta\, d\phi$$

$$N_{sc}(l) = \frac{\pi}{(k)^2}(2l+1)|1-\eta_l|^2\, v \qquad (6.3)$$

Partial scattering cross section is then

$$\sigma_{sc,l} = \frac{No\ of\ particles\ scattered\ with\ l\ value}{No\ of\ particles\ per\ unit\ area\ per\ \sec\ in\ incident\ beam} \qquad (6.4)$$

$$\sigma_{sc,l} = \frac{\pi}{(k)^2}(2l+1)|1-\eta_l|^2. \qquad (6.5)$$

Particle absorption cross section is calculated as

No. of particles absorbed $= \left|1-\left|n_l\right|^2\right|$. Therefore,

$$\sigma_{asc,l} = \frac{\pi}{k^2}(2l+1)\left[\left|1-\left|\eta_l\right|^2\right|\right]. \tag{6.6}$$

The above expression is valid for neutron scattering and absorption by a nucleus. Thus, one can calculate these cross sections if one determines the amplitude of the scattered wave.

6.16. Neutron Detectors

Neutrons have mass but no charge. Therefore, these do not produce ionization in matter, but these interact with nuclei by elastic or inelastic collisions and can be captured to produce γ rays.

The method of detection of neutrons and measurement of their energies depend upon their energy.

Low energy neutrons are captured by some nuclei such as boron, cadmium, and gadolinium and produce γ rays, which can be detected. One reaction used for detecting neutrons consists of a detector of boron-loaded liquid scintillator. The nuclear reaction is given as follows:

$B^{10} + n = Li^7 + \alpha$ particle.

The emitted α particle is detected, and its energy is measured by a scintillator detector. The Li^7 is left in an excited state of energy 0.48 MeV. Measurement of α particle energy gives the neutron energy.

The most commonly used method for the measurement of neutron energies is the time-of-flight method. A pulse of neutrons is produced in an accelerator. These neutrons travel a certain path of 10-200 meters. The neutrons are detected by a boron-loaded scintillator, and its time of arrival is measured. One can then determine the time of arrival measured from the instant the neutron pulse is produced. This data is then fed to a time-of-flight analyzer, which measures the time of arrival of neutrons and from the information of time one can determine its energy.

The relation between time and neutron energy for a 100 m flight path is $E(ev) = 72.3$ μsec/t. One eV energy neutron will take 72.3 μs to travel a distance of 100 meters.

A typical neutron time-of-flight spectra taken by Garg et al. at the Nevis Cyclotron Laboratory Columbia University, Physics Department is shown in figure 6.4.

6.17. Resonance Scattering and Reactions of Neutrons and Protons

In experimental study of neutron scattering and reactions, scientists observed that at some neutron or proton energies, the reaction cross section showed a large cross section (σ_r) at certain energies of incident particle known as resonant energy.

Total neutron cross sections on Zn[68] nucleus were measured by Garg et al. (6.6) using Oak Ridge electron accelerator and neutron time-of-flight detector. The part of the data in the neutron energy range of 40 to 90 keV is shown in figure 6.4a. The figure shows many sharp resonances at well-defined neutron energies. Neutron scattering resonances with l = 0 have asymmetrical shapes due to the interference between the hard sphere and the resonant amplitudes whereas neutron scattering due to l = 1 have small cross section and have symmetrical shapes.

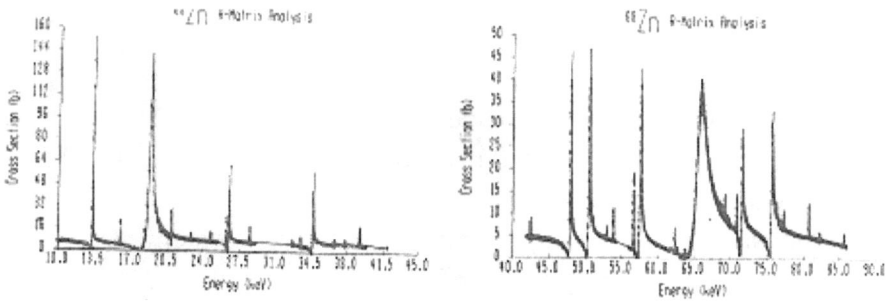

Figure 6.4a. Neutron time-of-flight spectra of total cross section of neutrons with Zn[68] nucleus in the energy range 10-90 keV

Measurements of neutron capture in the same nucleus Zn[68] taken by Garg et al. (6.6) in the energy range of 40 to 60 keV is shown in figure 6.4b. Here again, one sees many more resolved resonances. The reason is that capture cross section has the same capture widths for l = 0 and l = 1 resonances. Hence, one sees more p wave (l = 1) resonances in capture than in neutron scattering. From these measurements, one is able to determine probabilities of neutron elastic scattering and neutron capture.

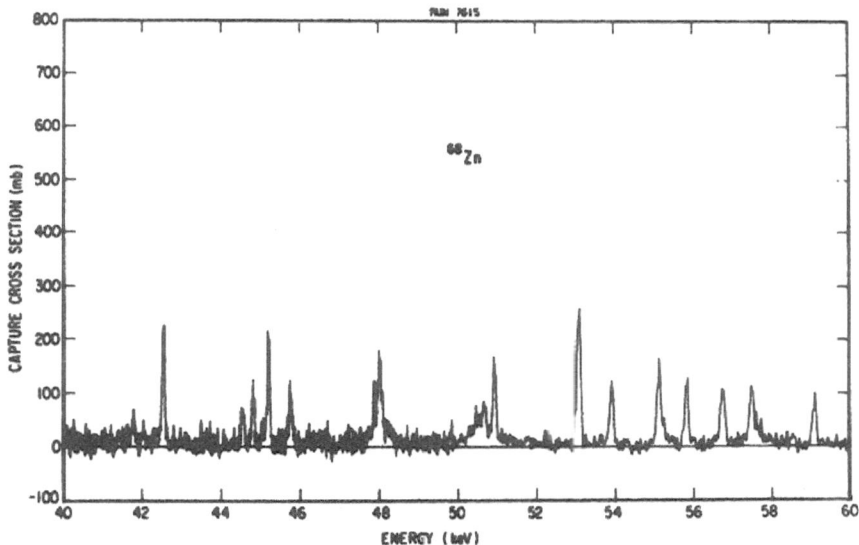

Figure 6.4b. Neutron capture cross section in Zn^{68}
in the energy range 40-60 keV

An explanation for the occurrence of resonances was given by N. Bohr (6.7). This is known as the compound nucleus theory. Bohr theorized that the incident neutron on entering the target nucleus was captured by the target nucleus and after its capture, it shared its energy with all the nucleons and formed a compound nucleus consisting of the neutron and the target nucleus. After a long time, compared to the transmission of the neutron through the nucleus, it decayed by the emission of the incident neutron (known as elastic scattering) or emission of gamma rays (known as capture) or emission of other particles (known as nuclear reaction). Bohr, in his compound nucleus theory, also postulated that the probability of decay was independent of the formation of the compound nucleus. (An important assumption made by Bohr, which was later verified by Ghoshal (6.8) as valid assumption.

Resonance Theory

A theory for the resonance scattering and reactions cross section can be developed as follows.

Let us consider neutron scattering at low neutron energy so that one can assume that only $l = 0$ neutrons are involved, and no Coulomb scattering is present, which will be the case for a charged particle reaction.

Neutron scattering is caused by the nuclear potential due to the nuclear force.

Let us consider that radial equation outside the range of nuclear potential for $r > R$ is given as follows:

$$\frac{d^2u}{dr^2} + k^2u_0 = 0, \ for \ r > R. \tag{6.7}$$

Solution of this equation gives

$$u_0(r) = c\sin(kr + \delta_o). \ for \ r > R \tag{6.8}$$

where δ_o = phase shift determined by the boundary condition and

$$k = \frac{\sqrt{2mE_n}}{\hbar} \ where \ E_n = neutron \ energy.$$

The term c is a constant to be determined.

First derivative of the solution (6.8) is

$$\left[\frac{du_0}{dr}\right]_{r=R} = \left[kc\cos(kr + \delta_0)\right]_{r=R} \tag{6.9}$$

We now define a quantity f_0 known as the logarithmic derivative, which is given as

$$f_0 = \left(\frac{r}{u_0}\frac{du_0}{dr}\right)_{r=R}$$

By combining equations 6.8 and 6.9, one gets f_0 for $r = R$ as

$$f_0 = kR\cot(kR + \delta_0) \tag{6.10}$$

where f_0 is positive but will decrease for increasing energy as more of the wave is packed into the potential well. It will go smoothly through zero, will become negative, and then will go to ∞ as $u(r)$ becomes zero at a large distance $r > R$.

If one considers incoming neutrons with $l = 0$, then the wave equation of incident neutron is given as

$$\psi = e^{ikz} = \frac{e^{ikr} - e^{-ikr}}{2ik}$$

and the radial wave function of scattered neutrons can be expressed as

$$u_0 = \frac{\sqrt{\pi}}{k} i \left| e^{-ikr} - \eta_0 e^{ikr} \right| \tag{6.11}$$

where η_0 is the amplitude of the scattered wave. Using the definition of f_0 as given earlier, we get for $l = 0$ the value of f_0 as

$$f_0 = \frac{e^{-ikR} \left(-ikR - \eta_0 (ik)e^{2ikR} \right)}{\left(1 - \eta_0 e^{2ikR} \right)} \tag{6.12}$$

Rearranging f_0 and η_0, one has the following:

$$\eta_0 = \frac{f_0 + ikR}{f_0 - ikR} e^{-2ikR}$$

Substituting the value of η_0 in equations 6.5 and 6.6 gives the elastic scattering and reaction cross section of neutrons with $l = 0$ as given below.

$$\sigma_{s,o} = \pi \lambda^2 \left(1 - \eta_0 \right|^2 \right) \qquad for\ \ l = 0$$

$$\sigma_\gamma = \pi \lambda^2 \left(1 - \left| \eta_0 \right|^2 \right) \qquad for\ \ l = 0$$

$$\left| 1 - \eta_0 \right|^2 = \left[1 - \frac{f_0 + ikR}{f_0 - ikR} e^{-2ikR} \right]^2$$

$$\left|1-\eta_0\right|^2 = \left[e^{2ikr} -1 +1 - \frac{f_0 + ikR}{f_0 - ikR}\right]^2 \tag{6.13}$$

$$\left|1-\eta_0\right|^2 = \left[e^{2ikr} -1 + \frac{-2ikR}{f_0 - ikR}\right]^2$$

$$\sigma_{s,o} = \pi\lambda^2 \left|e^{2ikr} -1 + \frac{-2ikR}{f_0 - ikR}\right|^2$$

If one defines potential and resonant amplitudes as

$$A_{pot} = e^{zikr} -1 \quad and \quad A_{res} = \frac{-2ikR}{f_0 - ikR} \tag{6.14}$$

The scattering cross section for $l=0$ is given as

$$\sigma_{s,o} = \pi\lambda^2 \left|A_{pot} + A_{\mathrm{Res}}\right|^2 \tag{6.15}$$

In general, one does not know the behavior of f_0 with high degree of accuracy, but its behavio r will be the same for other type of potential wells as for the square-well potential. Since f_0 goes smoothly through zero at a given neutron energy $E_n = E_s$ resonance energy, one can expand f_0 in a limited energy region around E_s as a power series.

$$f_o = -a\left(E - E_s\right)\left(\frac{\partial f_0}{\partial E}\right)_{E=E_s} + higher\ terms$$

By substituting this value of f_0 in equation 6.14 for the resonant amplitude and after neglecting higher terms in the expression, one gets

$$A_{res} = \frac{-2ikR}{\left(\dfrac{\partial f_o}{\partial E}\right)_{E=E_0} \left(E - E_s\right) - ikR} \tag{6.16}$$

If one takes the square of this amplitude A_{res} given in 6.16, it gives the behavior of neutron cross section near resonance energy E_s as in the following:

$$\sigma_{s,o} = \pi\lambda^2 A_{res}^2 = \pi\lambda^2 \left[\frac{-2ikR}{\left(\frac{df_0}{dE}\right)_{E=E_s}\left\{(E-E_s) - \frac{ikR}{\left(\frac{df_0}{dE}\right)}\right\}} \right]^2 \quad (6.17)$$

Defining neutron scattering width as $\dfrac{-2kR}{\left(\dfrac{df_0}{\partial E}\right)_{E=E_s}} = \Gamma_n$,

then $A_{res} = \dfrac{i\Gamma_n}{(E-E_s)-i\left(\dfrac{\Gamma_n}{2}\right)}$ \quad (6.18)

According to equation 6.14,

$$|A_{pot}| = e^{2ikR} - 1$$
$$= e^{ikR}\left(e^{ikR} - e^{-ikR}\right)$$

$$|A_{pot}|^2 = \left[2ie^{ikR}\sin kR\right]^2 \quad Since\ \sin kR = \frac{e^{ikR}-e^{-ikr}}{2i} \quad (6.19)$$
$$\sigma_{pot} = 4\pi R^2$$

Since for small values of kR $Sin kR = kR$. This cross section is known as hard sphere or potential scattering cross section.

Neutron-scattering cross section is square of the sum of the resonant and potential scattering amplitudes. Hence, scattering cross section at resonance energy is given by equation 6.15.

If $I^\pi = 0$ is the intrinsic spin and parity of the target nucleus, then J of compound nucleus state is obtained for $l = 0$ as

$$J = I \pm \frac{1}{2} \quad or \quad I - \frac{1}{2} \text{ and } I + \frac{1}{2}$$

Since neutron spin, $s = \frac{1}{2}$, one can obtain value of g known as the statistical weight factor, giving the percentage probability for the two states.

$$g(s) = \frac{2J+1}{(2S+1)(2I+1)} = \frac{2J+1}{2(2I+1)} \tag{6.20}$$

The value of g should be inserted in the cross sections. Scattering cross section σ_s at resonant energy is given as follows:

$$\sigma_{s,o} = g(s)\pi\lambda^2 \left[\frac{i\Gamma_n^s}{(E - E_s) + \frac{1}{2}i\Gamma_n^s} + 2kR \right]^2 + (1-g)4\pi R_x^2 \tag{6.21}$$

Reaction cross section σ_r is given as where Γ_r is the reaction width.

$$\sigma_{r,o} = g(s)\pi\lambda^2 \frac{\Gamma_n^s \Gamma_r^s}{(E - E_s)^2 + \frac{1}{4}\Gamma_n^2} \tag{6.22}$$

If the target nucleus has $I = 0$, the compound nucleus will have $J = \frac{1}{2}$, and the value of $g = 1$.

Whereas for intrinsic spin $I = \frac{1}{2}$ $\qquad J = \frac{1}{2} \pm \frac{1}{2} = 1, 0$

For

$$J = 1 \quad g_1 = \frac{2 \times 1 + 1}{2\left(2 \times \frac{1}{2} + 1\right)} = \frac{3}{4}$$

$$J = 0 \quad g_2 = \frac{1}{4}$$

Equations 6.21 and 6.22 are known as Breit-Wigner formula for resonance scattering and reaction cross sections.

Equation 6.21 has three terms.

1. Resonant scattering amplitude
2. Potential or hard sphere scattering amplitude
3. Interference between resonance and potential scattering amplitudes

If one neglects the potential scattering term, which is the case for the scattering of $l = 1$ neutrons, one has resonance scattering cross section symmetrical about the resonance energy whereas for $l = 0$ scattering, the resonance curve shows an assymetrical shape due to the interference of potential and resonant scattering amplitudes as shown in figure 6.5.

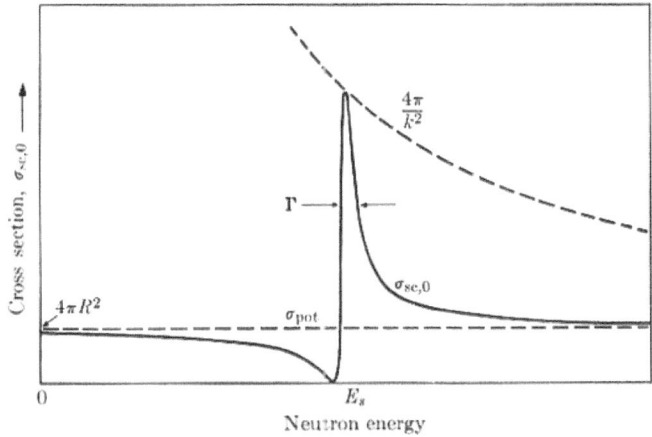

Figure 6.5 Elastic scattering cross section for $l = 0$ neutrons near a resonance in the compound nucleus. The angular momentum of the target nucleus is $J = 0$ and the reaction width $\Gamma_{r,0} = 0$, in which case the maximum scattering cross section $\sigma_{sc,0} = 4\pi/k^2$.

6.18. Formal Theory of Nuclear Reaction

After the observation of resonances in nuclear reaction and the theory developed by N. Bohr, Breit and Wigner derived a formula (6.9) for describing the cross section at a resonance. Many scientists proposed other general theories of resonance reaction. Among them were theories of Kapur and Peierls (6.10) and Wigner and Eisenbud (6.11).

These theories start with the incident particle moving in a potential well assumed to be a square well with $V(r) = 0$ for $r > R$. The resulting radial wave equation is

$$-\frac{\hbar^2}{2m}\frac{d^2u}{dr^2} + V(r)u = Eu .$$

The solution of the radial equation for $r = a$ gives rise to a number of virtual states in the potential well, but these states decay via different channels. The total cross section is given in terms of a collision function. The collision function is expressed in terms of an R- function, which is given as follows:

$$R_J = \sum_\lambda \frac{\gamma^2_{\lambda Jn}}{E_{\lambda J} - E - \frac{1}{2} + \Gamma_{\lambda\gamma}} \tag{6.23}$$

where the sum is over all levels of angular momentum J with the energy values E_λ and reduced neutron widths $\gamma^2_{\lambda Jn}$.

In the Breit-Wigner formula, the resonant cross section is described in terms of neutron and total widths and resonant energy. No consideration is given to the contribution to the cross section at resonant energy from neighboring resonances and from faraway resonances. The R-matrix theory takes into account the contribution from neighboring resonances to the resonance in consideration.

Neutron cross section measurements in nuclei all across the periodic table have been made by scientists in laboratories where high energy resolution techniques involving time of flight have been developed. Extensive measurements of total neutron cross section by Garg et al. (6.12) in U^{238} and Th^{232} nuclei in the energy range of few eV to about 1 MeV and many other nuclei were analyzed using the R-matrix formalism as well as an area analysis method based upon Breit-Wigner theory with correction for Doppler broadening of resonances. In general, the R-matrix shape analysis method is very successful for resonance reaction in light nuclei, and area analysis was useful for the analysis of resonances in heavy nuclei.

From these measurements, one can obtain values of resonance energies, neutron, and radiation widths and l-values of interacting neutrons. According to Bohr's theory, a resonance occurs when the resonance energy plus the neutron binding energy in the compound nucleus equals the well-defined excitation energy of the compound nucleus. Extensive measurements of these cross sections in nuclei have provided a large number of excited states with well-defined J and ℓ values. The established nuclear models are unable theoretically to reproduce energies of these highly excited states. The success of shell model lies in predicting the properties of only low-lying excited states.

One way to describe the general pattern of such highly excited states have been to investigate the level spacing D and their distribution about the mean value $<D>$. Similarly, one can investigate the distribution of neutron and radiation widths about their mean values.

6.19. Level-spacing Distribution

The investigation of the distribution of level spacing was initiated by E. Wigner in 1957 (6.13). Although the theoretical development did not take place until much later by Mehta and Gaudin and Mehta and Dyson (6.14). Wigner surmised a formula for level-spacing distribution as follows:

$$p(x)dx = \frac{\pi x}{2} e^{-\frac{\pi x^2}{4}} \tag{6.24}$$

where $x = D/\langle D \rangle$, <D> is the average level spacing.

Experimental measurement of the next nearest spacing distribution in U^{238} investigated by Garg et al. (6.12) is shown in figure 6.6.

The theoretical solid curve based upon Wigner's surmise is shown along with a curve based upon random distribution of level spacing. The data supports Wigner distribution, showing a very small probability for small spacing, meaning that the nuclear levels at high excitation energies repel each other.

Similar measurements of level spacing in many other nuclei from neutron-scattering cross section by Garg et al. (6.15) as well as proton

scattering cross section by Bilpuch et al.(6.16) have shown similar results
of level repulsion effects.

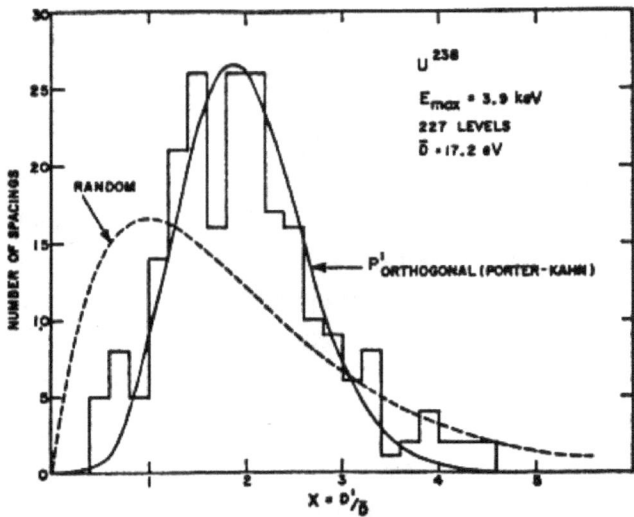

Figure (6.6) Histogram of the observed distribution of next-nearest-neighbor level spacings $x = D^1\langle D\rangle$ for U²³⁸. The theoretical curves shown for comparison

6.20. Neutron Scattering Width Distribution

The decay of the compound nucleus after its formation gives rise to the emission of a neutron or γ-rays. The probability of such emissions of neutron or gamma ray is expressed as Γ_n, Γ_r neutron and reaction widths, respectively.

Extensive measurements of high resolution neutron scattering and capture cross section have provided a large sample of neutron widths of levels of a given spin and parity.

Theoretical treatment of distribution of neutron-reduced widths about its mean value had been given by Porter and Thomas (6.17). Their expression is given as follows:

$$p(x)dx = \left(\frac{2}{\pi}\right)^{\frac{1}{2}} \exp\left(-\frac{X^2}{2}\right)dx \qquad (6.25)$$

where $x = \left(\Gamma_n^0 / \langle\Gamma_n^0\rangle\right)^{\frac{1}{2}}$.

The experimental data of 56 neutron-reduced widths for $l=0$ and 80 $l=1$ resonances in Zn^{68} taken by Garg et al. (6.6) is shown in figure 6.7.

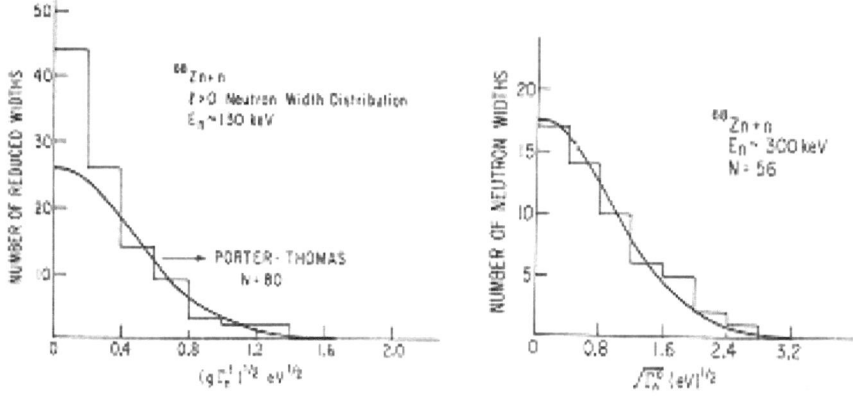

Figure 6.7. Neutron-reduced widths distribution and their comparison with Porter-Thomas distribution

It is clear that the experimental data within experimental errors agree with the prediction of Porter and Thomas distribution.

Similar behavior of capture and fission widths of neutron resonances have been reported by many authors.

6.21. Bohr's Assumption of Compound Nucleus Theory

Bohr made three assumptions in the development of his resonance theory.

1. Formation of compound nucleus after the capture of the incident particle.
2. Decay of compound nucleus after certain time emitting the captured particle.
3. Bohr postulated that the decay of compound nucleus was independent of its formation.

One can test the later assumption as follows: If α, β are the incident and outgoing channels, then the cross section for compound nucleus formation is given as follows:

$\sigma(\alpha, \beta) = \sigma_c(E_c, \alpha) G_c(E_c, \beta)$,

$\sigma_c(\alpha)$ = cross section for formation of CN at excitation energy E_c and α channel,

$G_c(\beta)$ = relative probability of decay by β channel.

There could be many outgoing channels after the capture of incident particle.

Therefore, if it can be shown that a compound nucleus that is formed at a given energy by the capture of different particles, then the decay of the compound nucleus will be the same. This will provide proof of Bohr's assumption.

S. N. Ghoshal (6.8) performed such an experiment. He used different particles, protons, and α particles and bombarded different nuclei such as Cu^{63}, Ni^{60} to form the compound nucleus Zn_{30}^{64} at the same energy of excitation and then measured the cross section for reactions (p, n), $(p, 2n)$, (p, p, n), (α, n), $(\alpha, 2n)$ and (α, pn), etc., as shown below.

If the decay is independent of the mode of formation, the cross sections obtained for different reactions for the two cases should be the same or

$\sigma(p, n) = \sigma(\alpha, n)$

$\sigma(p, 2n) = \sigma(\alpha, 2n)$

$\sigma(p, pn) = \sigma(\alpha, p, n)$

He found this was the case thereby confirming the validity of the independence assumption of Bohr's theory.

6.22. Continuum Theory

The compound nucleus theory is valid for particle energies less than few MeV. As the energy of the particles increases, the number of levels available to the incident particle after its capture becomes large since level density at excitation energy E is given as

$$\rho(E) = C \exp(2aE^{1/2})$$

where a is a constant and ρ = level density or number of levels per unit energy. The formula shows that the level density increases exponentially as a function of excitation energy of the nucleus.

The capture of a neutron leads to overlapping of levels at the binding energy of about 8 MeV, and the decay of the nucleus follows a statistical theory.

One can thus learn about the level densities of nuclei from the measurements of resonances due to capture of neutrons. According to Fermi gas model, the level density is given as

$$\rho(E) = \rho(E) \, (2J + 1) \exp \left(-(J + 1)^2/2\sigma^2\right) \tag{6.26}$$

where J is the total angular momentum of the excited states, and σ is the cutoff factor. The formula shows that the level density depends upon the J values.

Neutron cross section measurements in V, Co, and Mn nuclei by Garg et al. giving values of level densities for two different values of J shows that the level density is independent of J values. From these measurements, they determined a value of spin cutoff factor $\sigma = 3$.

6.23. Charged Particle Nuclear Reactions

Nuclear reactions with charged particle such as protons, deuterons, and alpha particles have been studied extensively. As mentioned earlier, charged particles require enough energy to penetrate the nucleus due to the Coulomb repulsion.

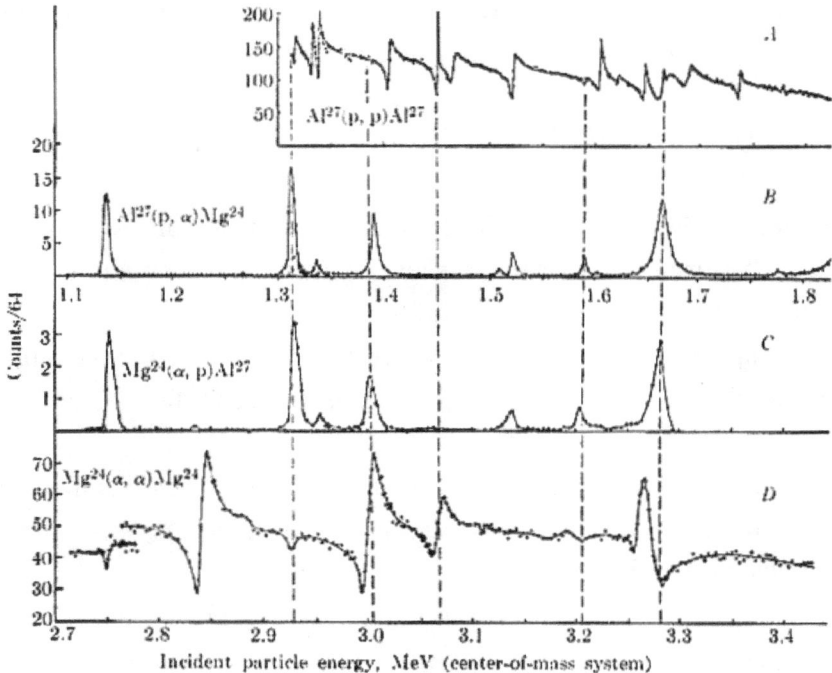

Figure 6.8   Excitation curves for the $Al^{27} + p$ and $Mg^{24} + \alpha$ systems displaying several resonances in the compound nucleus Si^{28}. [From Kaufmann *et al., Phys. Rev.* 88, 675 (1952).]

A graph showing the height of Coulomb barrier for these particles was given in figure 6.3. Even for a light nucleus such as C^{12}, the Coulomb barrier for protons is about 1 MeV, which means that the minimum proton energy needed for causing a nuclear reaction is in excess of this energy. Such energies can be produced using a Van de Graaff or a tandem accelerator. Many types of detectors such as solid state or scintillation detectors are used for the detection and measurement of energies of charged particles.

When protons of a given energy strike a target nucleus, the most prolific reaction is the elastic scattering of protons (p, p). If the energy of the protons is greater than the excited state energy of the nucleus, inelastic scattering of protons can compete with elastic scattering. In addition, the proton can be captured and emit gamma rays known as the capture reaction.

Nuclear reactions with deuterons, tritons, He^3, and He^4 particles are usually direct reactions without the formation of the compound nucleus.

An example of nuclear reaction studied by Kaufman (6.18) with protons and alpha particles is shown in figure 6.8.

The reactions studied are elastic scattering of protons and (p, α) reactions on Al^{27} in the energy range of 1.3 and 1.8 MeV. The data shows many resonances in each reaction. In the proton elastic scattering $l = 0$, resonances show asymmetrical shapes as is observed in neutron elastic scattering.

In the α particle-induced reactions on Mg^{24} in the energy range from 2.7 to 3.4 MeV, one observes many resonances with pronounced asymmetrical shapes for $l = 0$ resonances and symmetrical shapes for $l = 1$ resonances.

The analysis of these reactions by using R-matrix formalism provided information about the energies and J^π values of excited states of the compound nucleus.

Charged particles' reactions are limited to light nuclei in view of the Coulomb barrier that the particles have to overcome whereas neutron can interact with almost zero energy.

6.24. The Optical Model

Extensive measurements of average total neutron cross section by Barshall et al. (6.19) in many nuclei as a function of neutron energies up to few MeV showed broad resonances at certain mass number. Since the measurements were made at few MeV, individual resonances are not separated in view of poor resolution of the experiment and showed only the gross structure of the resonant cross section.

Figure 6.10 shows a three-dimensional plot of total neutron average cross section for nuclei as a function of atomic weight and as a function of neutron energy throughout the periodic table of elements. This figure was compiled by Feshbach et al. (6.20) from the data of Barshall et al. (6.19).

Wide resonances are seen at certain neutron energies and at certain A values. Total cross section in general decreases with increasing energy.

The l values of partial waves that are responsible for resonances are indicated in the figure as s, p, d for l = 0, 1, and 2, respectively. Broad resonances are observed for each l value at different atomic weights.

Such broad resonance were interpreted by Feshbach, Porter, and Weisskopf (6.20) as produced by a complex nuclear potential having a real part and an imaginary part called the optical potential in analogy with behavior of light waves being reflected and transmitted by a piece of glass.

In the optical model also known as cloudy crystal ball model, Feshbach et al. proposed that the nucleus is translucent to particle waves. The absorption and scattering of the particle by the nucleus is treated by assuming that the nucleus has a complex potential such as $V = -(V + iW)$.

Both potentials are assumed to depend upon the distance r from the center of the nucleus and in general has the Woods-Saxon form. The terms V and W are positive real numbers.

Using this form of the potential, one can write the Schrödinger wave equation for the neutron scattering of energy E as in the following:

$$\frac{d^2\psi}{dx^2} + \frac{2m}{\hbar^2}\left(E + V_0 + iW\right)\psi = 0 \qquad (6.27)$$

The plane wave solution of the above equation is

$$\psi = e^{\pm ik_c x} \qquad (6.28)$$

where the wave number K inside the potential is given as

$$K_c = \frac{1}{\hbar}\sqrt{2m\left(E + V_0 - iW\right)} \qquad (6.29)$$

where m is the reduced mass, E is the neutron energy, and $V_0 + iW$ is the complex potential whose values are to be determined from the experimental data.

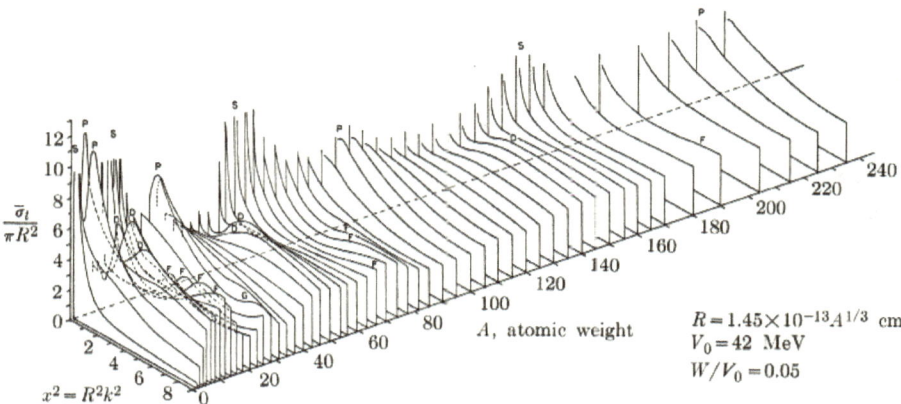

Figure 6.9. Calculated "gross structure" of total neutron cross section, as a function of the energy parameter x^2 and A. [From H. Feshbach, C. E. Porter, and V. F. Weisskopf, *Phys. Rev.* **96**, 448 (1954).]

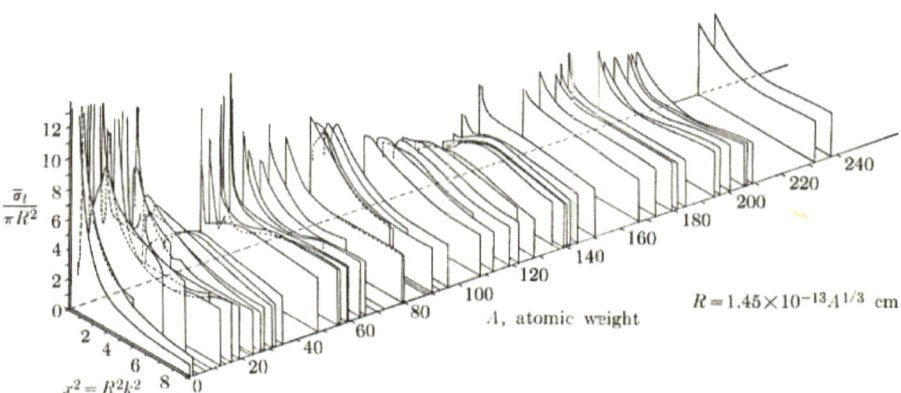

Figure 6.10. Observed "gross structure" of total neutron cross section. (Compiled by Feshbach *et al.*, *loc. cit.*)

In this model, the real part of the nuclear potential V_0 gives rise to scattering, and the imaginary part of the potential W gives rise to particle absorption. Thus, the wave number K_c can be separated into a real and imaginary wave number as follows:

$$K_c = K + ik \quad where \quad K = \frac{1}{\hbar}\sqrt{2m(E + V_0)}. \qquad (6.30)$$

Substituting in 6.28 above, we get

$$\psi = e^{\pm i(K+ik)x} = e^{iKx} + e^{-ikx} \tag{6.31}$$

This wave function represents an exponentially decaying wave with a mean decay length (L) known as mean free path.

$$L = \frac{1}{2k} = \frac{E+V_0}{WK} = \frac{E_0+V}{2\pi W}\lambda \tag{6.32}$$

$$\text{since } K = \frac{2\pi}{\lambda}.$$

As an example, consider the incident neutron energy is $E = 10$ MeV and optical potentials are $V_0 = 40$ MeV, and $W = 10$ MeV, one gets

$$K = 0.2187(50)^{\frac{1}{2}} fm^{-1} \tag{6.33}$$

$$K = 1.55 \, fm^{-1}$$

Hence, the mean free path for neutron of 10 MeV energy is

$$L = \frac{(10+40)\lambda}{2\pi (10)} = 0.80\lambda = 3.23 \text{ fm.} \tag{6.34}$$

This shows that the mean free path is comparable to the dimension of the nucleus. One can calculate the mean free path (L) of nucleons inside nuclei from the strength of the optical model potential as a function of particle energy.

Equation 6.31 represents a travelling wave whose amplitude is decreasing exponentially as it advances. It therefore represents a beam of particles, some of which are being absorbed.

Physical significance of the decaying wave function is that as far as the world outside the nucleus is concerned, the particle may disappear into the nucleus. What exactly goes on inside the nucleus is not correctly described by a decaying wave function. If the particle is a nucleon, it may collide with other nucleons and share its energy with them. Even if it has enough energy left to escape from the nucleus, it is no longer identifiable as a particle in the beam or as an elastically scattered particle.

The only way a particle can be removed from the incident beam is by having a collision with one of the nucleons inside the nucleus. This probability depends upon the (L), the mean free path between collisions.

Average time between collisions is the mean free path (L) divided by the particle velocity (v). Based upon the values of nuclear potential, one would expect maxima in cross sections whenever neutron energy (zero) agrees with one of the (l = 0) levels of the optical model potential. The mass (A) where these maxima will occur will be given by K and R values where

$$K = \frac{1}{\hbar}\sqrt{2m_n\left(E_n + V_0\right)} \quad and \quad R = r_0 A^{\frac{1}{3}}. \tag{6.35}$$

Assuming a value of V_0 = 40 MeV, E = 10 MeV, and r_0 = 1.25 fm

$$KR = \frac{1}{\hbar}\sqrt{2m_n\left(V_0 + E_n\right)R^2} \tag{6.36}$$

For $E_n << V_0$

$$KR = \frac{1}{\hbar}\sqrt{2m_n V_0 R^2} \tag{6.37}$$

For resonance

$$Cot\ KR = 0, or\ KR = (2n+1)\frac{\pi}{2}\ where\ n = 1, 2, 3$$

For n = 2, $KR = 2.5\pi$ and   for $n = 3$, $KR = \frac{7}{2}\pi$.

Substituting values of KR from 6.33, one gets A = 55 and A = 150.

These are atomic masses where maximum values of average cross section for l = 0 were observed by Barshall et al.

The initial application of optical model was made in describing the average total cross section of neutrons at various energies on a large number of nuclei.

Calculated total cross sections by Feshbach et al. using values of $V = 42$ MeV and $W = 2.1$ MeV (6.20) are plotted in figure 6.9(b). This figure shows maximum values of total cross section for different l-values labeled as $s, p,$ and d resonances for different masses. For example for $l = 0$, s-wave maxima occurs at $A = 50$ and at $A = 160$, and for $l = 1$, p-wave maxima occurs at $A = 30, 90,$ and 230.

6.25.a. Neutron Strength Function

Application of the optical model, which was successful in predicting the broad resonances observed in the neutron total average cross section, has also found success in predicting the occurrence of maxima in the neutron strength functions as a function of atomic weights as well as in predicting the behavior of angular distribution observed in direct reactions.

One of the quantities obtained from the measurement of neutron cross section is the neutron resonance widths (Γ_n). If one divides this width by the resonance energy, $\sqrt{E_o}$ are obtains a reduced neutron width $\Gamma_n^0 = \dfrac{\Gamma_n}{\sqrt{E_o}}$. If one sums these values of Γ_n^0 in a certain energy interval E and divides this by the average level spacing D, one then obtains a quantity known as strength function, which is defined as $S_0 = \left(\dfrac{\Gamma_n^0}{D} \right)$ for $l = 0$ resonances. Similarly for $l = 1$ resonances, one obtains the p-wave strength function.

From the measurements of neutron widths over a large energy range, scientists have obtained values of strength function for $l = 0$ and $l = 1$ neutrons.

A plot of s-wave ($l = 0$) strength function S_0 as a function of atomic mass A is shown in figure 6.11. The experimental data points are a compilation of measurements made by scientists at many laboratories such as Columbia University, Duke University, Argonne National Laboratory (ANL), Brookhaven National Laboratory (BNL), and Oak Ridge National Laboratory (ORNL).

In this figure, one sees maximum value in the S_0 value at $A = 55$ and $A = 155$. The peak near $A = 155$ splits in two peaks one at $A = 145$ and the other at $A = 185$. This split is due to the fact that nuclei in this region are deformed and require a modified treatment, taking consideration for collective behavior. The solid curve is based upon the optical model calculation using the following parameters $V_0 = 52$ MeV, $W_0 = 3.1$ MeV. The potential used had a Woods-Saxon form. This model was proposed by Feshbach et al.

The explanation for such peak values of strength functions at certain mass number is attributed to shell model structure of nuclei. According to the shell-model calculation based upon an assumed spherical potential well with spin orbit coupling, the $2s$ and $3s$ states occur at mass numbers 40 and 160 and $3p$ state occurs at $A = 80$. The maxima in the strength function are observed close to these atomic masses.

Similarly, a plot of strength functions for $l = 1$ neutron scattering in many nuclei measured by scientists in many laboratories mentioned in the figure caption is shown in figure 6.12. The figure shows maximum values at $A = 90$ and $A = 110$. Solid lines are the theoretical fit based upon the optical model calculation. These are the approximate values of A where maxima of strength functions are observed. The concept here is that the width of the single particle p-state is divided among a large number of resonances in a given energy interval.

6.25.b. Doorway States

The optical model was quite successful in explaining the broad resonances observed in the average neutron total cross section as well as broad resonances observed in neutron strength functions for $l = 0$ and $l = 1$ resonances.

Block and Feshbach (6.37) suggested that there may also exist resonance structure intermediate between the broad resonances observed in the neutron average cross section and the very fine resonances due to the formation of the compound nucleus.

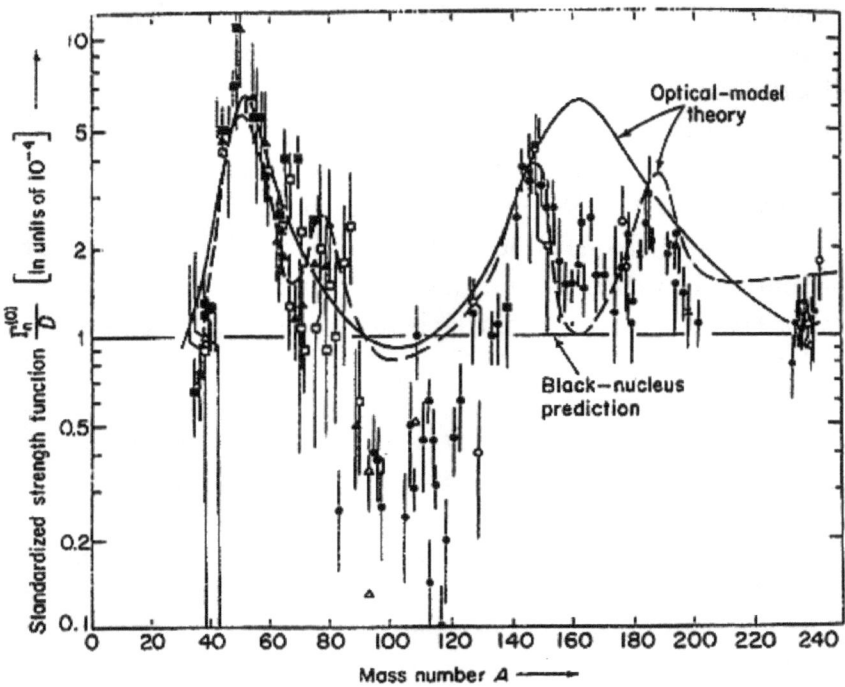

Figure 6.11. The s-wave neutron strength function versus atomic mass

According to them, when a neutron enters a nucleus, it interacts with a single nucleon before sharing its energy with many nucleons leading to the formation of the compound nucleus. The configuration resulting from first collision excites two particles—the incoming neutron and the other inside the nucleus thereby leaving a hole in its former position. This is called 2p-1h state also known as doorway state to the formation of the compound state. If 2p-1h state is long-lived, a sub-giant resonance much narrower than those attributed to single particle state might be observed in the average cross section near the energy of 2p-1h state.

In the extensive measurements of neutron cross section in many nuclei, one measures the widths of resonances and then sums these in a given energy interval. If one sums these widths in a smaller interval of energy and finds the sum to be very large in certain region, this may indicate the presence of such doorway states.

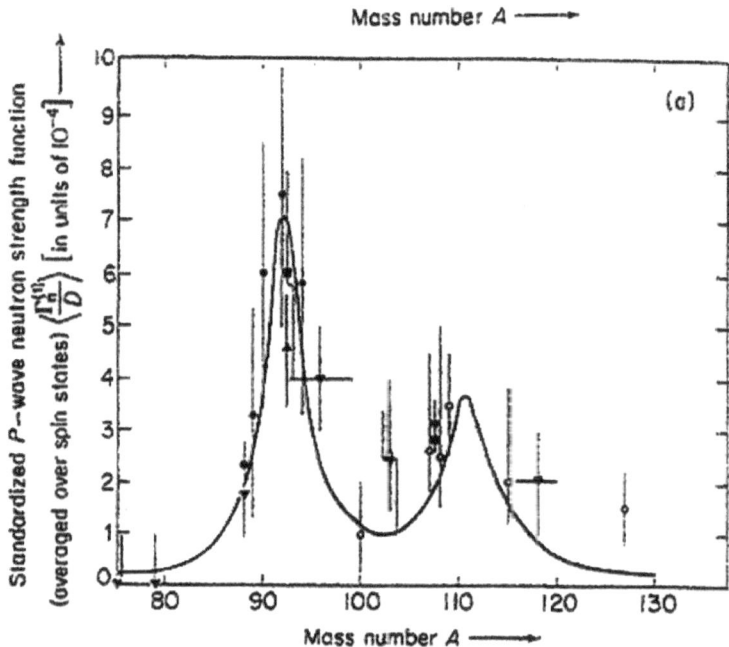

Figure (6.12) p-wave strength function vs atomic mass A. points are compilation of measurements of different groups cited in the text. Solid curve is the theoretical fits to the data.

The measurement of neutron cross section by Wilenzick et al. (6.38) in doubly magic nucleus Pb^{208} showed a broad s-wave resonance at the neutron energy of 500 keV with the neutron width of 58 keV, as shown in figure 6.12 (b) curve A.

Similar neutron cross section measurements in Pb^{206} and Pb^{207} were performed by Farrel et al. (6.21). Their results are plotted in figure 6.12(b) curve B. This figure shows that there are many resonances grouped together at the neutron energy of 500 keV with the sum of neutron widths of the same value as was observed in the case of Pb^{208}.

Therefore, one can conclude that a 2p-1h state seen as a single resonance in Pb^{208} acts as sub-giant resonance in the cross section observed in Pb^{206}.

Similar doorway states have been observed in many other nuclei.

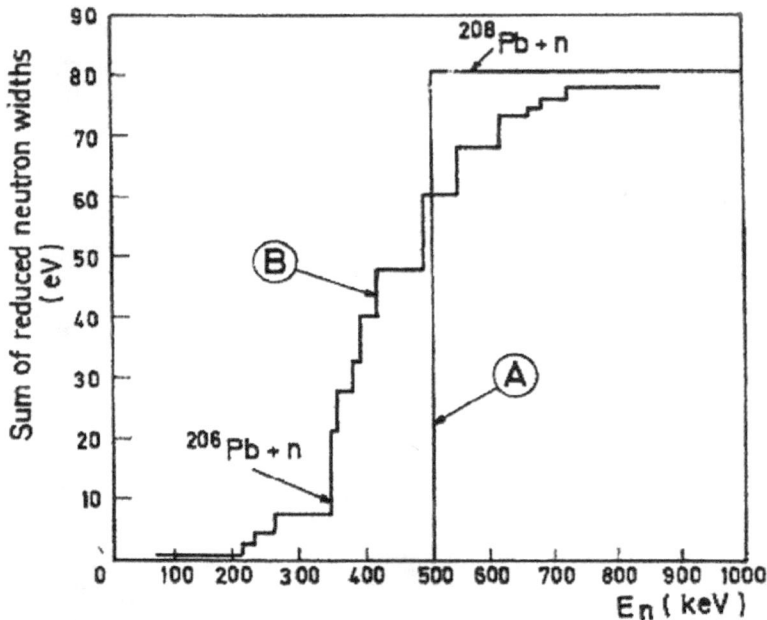

Figure 6.12(b). *A* Neutron resonance in Pb^{208} at E_n = 500keV *B* Sum of *s*-wave neutron reduced widths in Pb^{206} as a function of neutron energy

6.26. Direct Reactions of Particles with Nuclei

Another class of nuclear reactions is known as direct reactions where the incident particle interacts for a short time with only a single nucleon inside the nucleus.

Direct reactions include inelastic scattering, stripping reactions, knockout, and pickup reactions.

In knockout reaction, an incident particle such as a neutron will knock out another neutron or a single proton from inside the nucleus, and the incident particle will be captured by the target nucleus. Such reactions display an angular distribution of outgoing particle, which is forward peaked. Such a reaction is fast and does not undergo formation of compound nucleus as is the case for resonance reaction, which has a lifetime of about 10^{-15} seconds as opposed to direct reaction with a lifetime of about 10^{-22} seconds.

Another class of direct nuclear reactions is stripping of incident composite particle such as deuteron, triton, He^3 incident particles.

In this type of reaction, incident deuteron or a composite particle, which is a loosely bound nucleus is split into a neutron and a proton; one of these particles is captured by the target nucleus, and the other particle goes on in the forward direction as the outgoing particle. The inverse of stripping reaction is known as pickup reaction in which the incident particle might pick up a neutron, a proton, or a composite particle from inside the nucleus to form a deuteron, triton He^3 or a He^4 nucleus, which is emitted from the target nucleus.

6.27. Stripping Reactions

As an example of this type of reaction, deuteron stripping reactions such as (d, p) and (d, n) will be discussed. As mentioned earlier, stripping reactions can also take place with heavier composite particles.

Semiclassical description of the deuteron stripping reaction is shown below in figure 6.13.

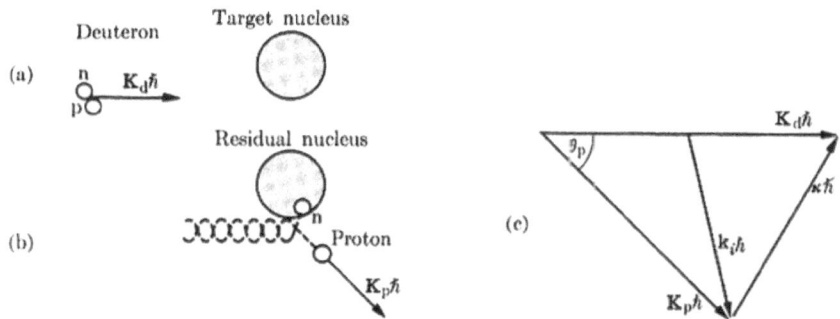

Figure (6.13) Deuteron stripping; (a) before the collision, and (b) after the reaction has taken place. (c) Diagram used to illustrate conservation of momentum.

In this reaction, a beam of deuterons, consisting of a proton and a neutron, moves from left toward the target nucleus, which is heavier than deuteron. One of the constituent of deuteron such as a neutron is captured by the target nucleus to form a residual nucleus in a specific state. The remaining proton moves away from the target nucleus in a forward direction with a momentum $k_p \hbar$. The term $k\hbar$ is the angular momentum transferred to the residual nucleus.

When $k_d \hbar$, $k_p \hbar$, and θ_p are known, $k\hbar$ can be determined by the cosine rule applied to momentum triangle shown above in figure 6.13 and is given as

$$(k\hbar)^2 = \left(k_p \hbar\right)^2 + \left(k_d \hbar\right)^2 - 2k_p k_d \hbar^2 \cos\theta \qquad (6.38)$$

As a simplification, the assumption is made that outgoing proton does not come within the range of nuclear force or Coulomb force. Linear momentum $k\hbar$ imparted to the nucleus is the momentum brought to the residual nucleus by the neutron.

The orbital angular momentum is given as $l = rxp$, where r is the impact parameter equal to the nuclear radius (R) and p is the linear momentum of the captured neutron.

Orbital angular momentum (l) is quantized and has a value of $\sqrt{l(l+1)}$.

The strong outcome of the above discussion is that the formula predicts the angular distribution of outgoing particle. Experimentally, one studies the angular distribution of the emitted particle that is proton in this case, and from these measurements, one can determine the l-value of the neutron captured by the residual nucleus.

Oppenheimer and Philips (6.22) studied low energy stripping reactions and observed that the angular distribution of the emitted particle was predominantly in the forward direction for small angles θ.

Theoretical work on stripping reactions by Butler (6.23) showed that the reaction was a single-step process and the intensity of the outgoing particle as a function of angle Θ is given by the square of spherical Bessel function of order l as

$$I(\theta) \propto \left[j_l \left(qR\right) \right]^2 \qquad (6.39)$$

where $I(\theta)$ is the intensity of protons as a function of the angle Θ, j_l is the spherical Bessel function of order l. Values of spherical function are available from tables. In general, spherical Bessel function for $l = 0$ shows a maximum at 0^0 degree angle and gives maximum at small angles for $l = 1$. The maximum shifts to larger angles for $l > 1$.

As an example of (d, p) reaction, one bombards a target nucleus scandium Sc^{45} with deuterons of Energy E = 6.074 MeV. Figure 6.14 shows the spectra of protons emitted leading to the ground state and excited states of residual nucleus Sc^{46}.

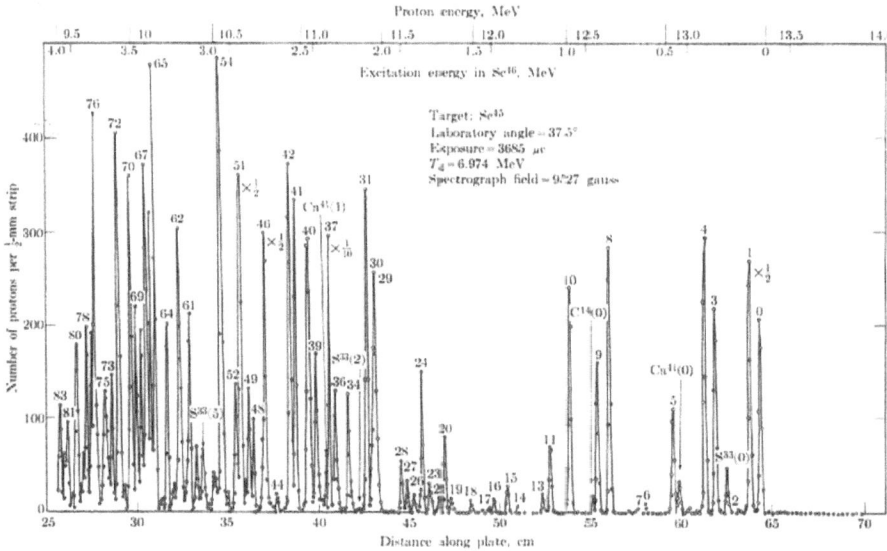

Figure 6.14. Proton spectra from Sc ^{45} (d, p) Sc ^{46} reaction

The reaction is written as $Sc^{45}(d, p)Sc^{46*}$. Such spectra is taken at different angles from $0°$ to $150°$, and the intensity is plotted as a function of θ. Intensity represents differential scattering cross section $\dfrac{d\sigma}{d\Omega}$ expressed in mb/sterad.

This spectrum was taken by Rapaport et al. (6.24) at E_d = 6.974 MeV with a multigap spectrograph. It shows protons of different energies with several intensities.

Selected plots of differential stripping cross section for protons leading to levels labeled as 1, 37, and 48 in figure 6.14 are shown in figure 6.15.

Experimental points are compared with the theoretical calculation (solid curve), which are based upon the distorted wave Born approximation theory known as DWBA.

It is shown that angular distribution of the ground state (level 1) fits
with $l_n = 3$ and to excited states level 37 with $l_n = 1$ and level 48 with $l_n = 0$.
Knowing the value of angular momentum (l) of neutron, one can determine
the J^π value of ground state and excited states of the residual nucleus, Sc^{46}.

If I_i and I_f are the spins of the initial and final states of the target and
residual nuclei respectively, then conservation of angular momentum
gives the following:

$$I_i + I_f + \frac{1}{2} \geq l_n \geq I_i + I_f - \frac{1}{2}$$

where l_n is the angular momentum of captured neutron. Law of conservation
of parity determines whether the l-value is even or odd.

If l-value is ascertained from the angular distribution measurement,
one can determine the parity and spin of the final nucleus if unknown.

Thus, such stripping reactions offer an excellent spectroscopic tool
for the determination of J^π of states of nuclei.

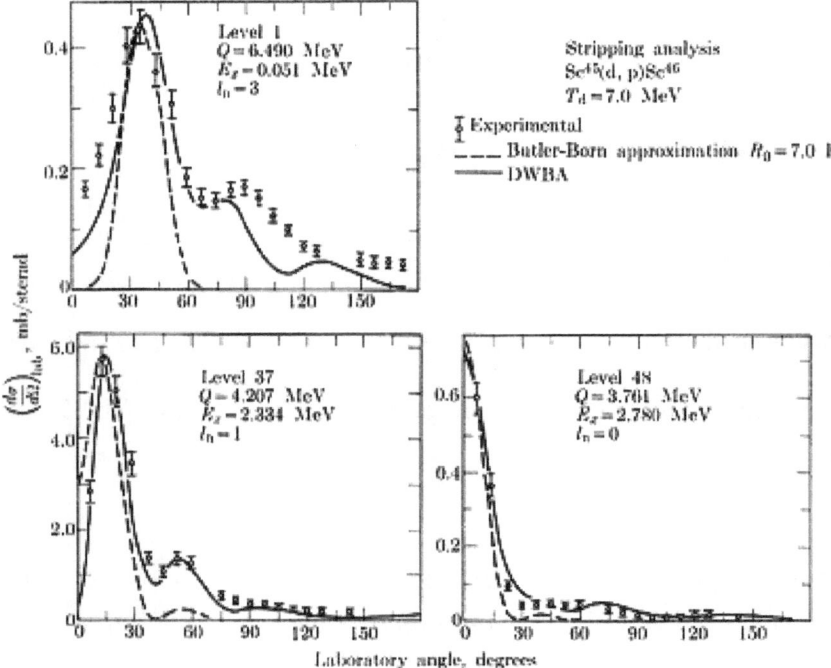

Figure (6.15) Experimental stripping cross sections compared with distorted-wave
born approximation, (DWBA) calculations as well as plane-wave (Butler) calculations.
(From J. Rapaport, A. Sperduto, and W. W. Buechner **Phy Rev 151,939 (1966)**

6.28. Theories of Stripping Reaction

Theories of stripping reactions were developed by many scientists in particular by Butler (6.23), Tobocman (6.25), Huby et al. (6.26), and Bhatia et al. (6.27) independently.

In general, direct reaction theories are known as plane wave Born approximation (PWBA) developed by Butler and the theories of distorted wave Born approximation (DWBA) developed by Tobocman (6.25) and Huby et al (6.26). In PWBA, the incident and outgoing particle waves are not affected by either the strong nuclear or Coulomb potentials. Such a simple theory was found to be surprisingly successful as a first approximation. In DWBA theory, consideration is given for strong interaction of target nucleus with the incident and outgoing particles by using a complex optical potential as provided in the optical model. The incident and outgoing particles are treated as plane wave of the form e^{-ikr}, and they use the plane wave Born approximation or a refinement, which takes into account the distortion of outgoing particle wave through an interaction between particle and target, which yields the distorted wave born approximation formalism. The latter theory is widely used in analyzing the experimental data of direct reactions.

Direct stripping reactions are essentially processes that occur in the surface region of the nucleus where the incident and outgoing waves of particles are not greatly affected by the strong nuclear potential.

The theory also assumes that the captured neutron in (d, p) reaction goes into a single particle shell-model state.

It was pointed out by Bethe and Butler (6.28) that stripping reactions in some cases could be used to measure the parity of shell model states. An example of this determination is the stripping reaction $P^{31}(d, p)P^{32}$.

In this reaction, the target nucleus P^{31} has $J = \frac{1}{2}$. According to the shell model, the odd proton is in $2S_{1/2}$ orbit. In (d, p) reaction, a neutron is captured to form P^{32}. According to shell model, the neutron will go in $d_{3/2}$ orbit. The ground state of P^{32} is 1^+. Hence, this state can be reached from $\frac{1}{2}+$ by $l = 1$ or $l = 2$. However, the maximum cross section for this reaction fits with $l = 2$ consistent with the shell-model prediction.

Extensive studies of such reaction in many nuclei has established the validity of DWBA theory, which takes into account the nuclear and

Coulomb interaction of residual nucleus with the incident and emitted particles.

Theoretical treatments of knock out reaction in which the incident composite particle knocks out a particle from inside the nucleus and pick up reactions where the incident particle such as a proton picks up a neutron from inside the nucleus forming a deuteron which is the outgoing particle follow a similar approach. Analysis of these data use direct reaction theories of PWBA or DWBA.

6.29. Proton inelastic Scattering Reaction

In these types of reactions the incident proton on entering the target nucleus imparts part of it's energy to a nucleon inside the target nucleus and is emitted with energy less than the incident energy. The energy difference is used to excite the nucleus to one of the known excited state, which emits a gamma ray of known energy. Inelastic scattering reaction is considered as one step process indicating direct reaction mechanism

An example of such a reaction is the measurement by Kikuchi et al (6.29). of inelastic scattering of 14.1 Mev protons in the target Cr ^{52} leaving the residual nucleus in 1.45 Mev excited state. This reaction is shown in the figure (6.16).

The shape of the angular distribution of protons is fitted with PWBA direct reaction theory. This is shown in the figure (6.16). Theoretical curve (solid line) based upon PWBA theory is evaluated for l=2 and R= 6.35 fm. The fit to the data for the main peak is excellent. The fit at large angles from 90 to 180 degrees can be improved by using DWBA theory.

Extensive measurements of inelastic scattering of 14Mev neutrons have also been made. All these measurements show a direct type of reaction, which can be fitted with PWBA theory.

This is shown in figure 6.16. Theoretical curve (solid line) based upon PWBA theory is evaluated for $l = 2$ and $R = 6.35$ fm. The fit to the data for the main peak is excellent. The fit at large angles from 90 to 180 degrees can be improved by using DWBA theory.

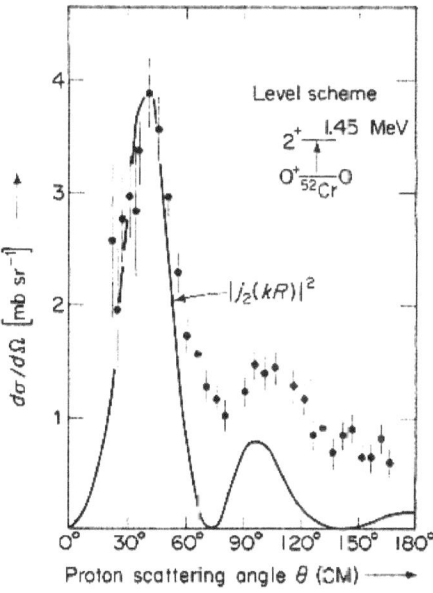

Figure (6.16) Angular-distribution results for the inelastic scattering of 14.1-MeV protons on chromium, (83.76 percent ^{52}Cr) (solid points) compared with the arbitrarily normalized $|j_2(kR)|^2$ curve (solid line) predicted by PWBA theory (the depicted fit ensued on setting R to 6.35 fm, which may be compared with the nuclear radius of ^{52}Cr. viz. $R = 1.4A^{1/3} = 5.23$ fm).

6.30. Other Nuclear Reactions

There are many other types of nuclear reactions, which have been studied extensively. An extensive list of such reactions cannot be given here, but some of the most common reaction studied are listed below.

6.31. (He^3, p), (He^3, n), and (He^3, d) Reactions

Nuclear reactions induced by low energy He^3 particles show that the angular distribution of outgoing neutron is isotropic. This suggests that the reaction mechanism is a compound nucleus formation and its subsequent decay. Whereas the nuclear reaction at higher energies above 4.0 MeV measured by Gale et al. (6.39) show that the angular distribution of emitted neutrons shows forward peaking characteristic of stripping reactions.

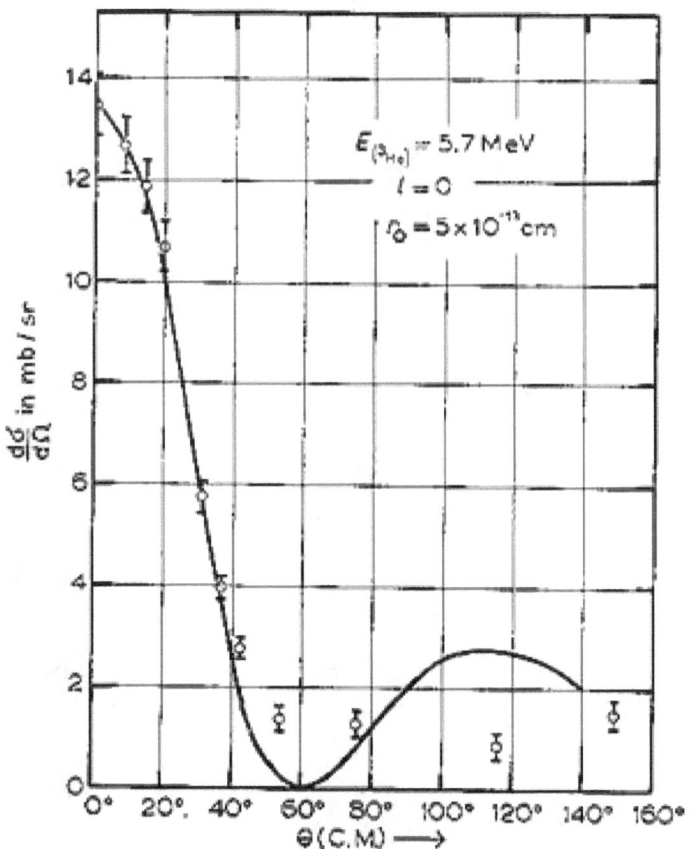

Figure 6.17 Angular Distribution of Neutrons
in the C^{12} (He^3, n) O^{14} Reaction

This behavior of angular distribution of neutrons is similar as the d, p reaction with nuclei. The He^3 particle has two protons and one neutron. Hence, in this reaction induced by He^3, two nucleons are stripped off from the third nucleon. In (He^3, n) reaction, diprotons are captured by the target nucleus, and the neutron is emitted with the energy available in the reaction.

This reaction is known as double stripping reaction. The theory of such reactions was developed by Newns (6.40). A typical neutron angular distribution of neutrons produced in C^{12} (He^3, n) O^{14} reaction leading to the ground state of oxygen taken by Gale et al. (6.39) at the energy of 5.70 MeV is shown in figure 6.17.

Solid curve in the figure is due to the stripping theory of Newns using spherical Bessel functions of order $l = 0$ and a nuclear radius of 5 fermis as discussed in section 6.27. Here $l = 0$ is the angular momentum of the captured diprotons transferred to the residual nucleus.

The (He^3, p) reactions studied by Forsyth et al. (6.41) at about the same energies as the (He^3, n) reactions show that the angular distribution of outgoing protons is forward peaked, characteristic of stripping reactions. The (He^3, n) reaction is also important because it produces proton rich residual nuclei whose properties could be of great interest.

6.32. The α Particle-induced Nuclear Reactions

The α particles have two units of charges; hence, these require much higher energies than protons and deuterons for a nuclear interaction with a target to overcome the Coulomb barrier. These α-particle reactions have been studied at energies between

10 and 40 MeV.

The α particle interaction has historical importance. Experiments of α particle scattering from gold nucleus was investigated by Geiger and Marsden and interpreted by Rutherford to establish the existence of the nucleus.

Around 1950, S. N. Ghoshal studied α-particle-induced reactions such as (α, n), (α, p), and $(\alpha, 2n)$ in Ni^{60} and compared them with the output of reactions' products induced by protons exciting the same compound nucleus to the same energy, and he determined the cross sections in both cases. He found the cross section to be the same, establishing the assumption made by Niels Bohr in his compound nucleus theory that decay of compound nucleus was independent of the mode of formation.

In general, α particles produce different types of nuclear reactions when these interact with nuclei. The most predominant reactions is elastic and inelastic scattering of α particles. The interaction of α particles with nuclei produce oscillatory behavior of the angular distribution of the scattered α particles as a function of scattering angle. Such experiments are performed by many physicists.

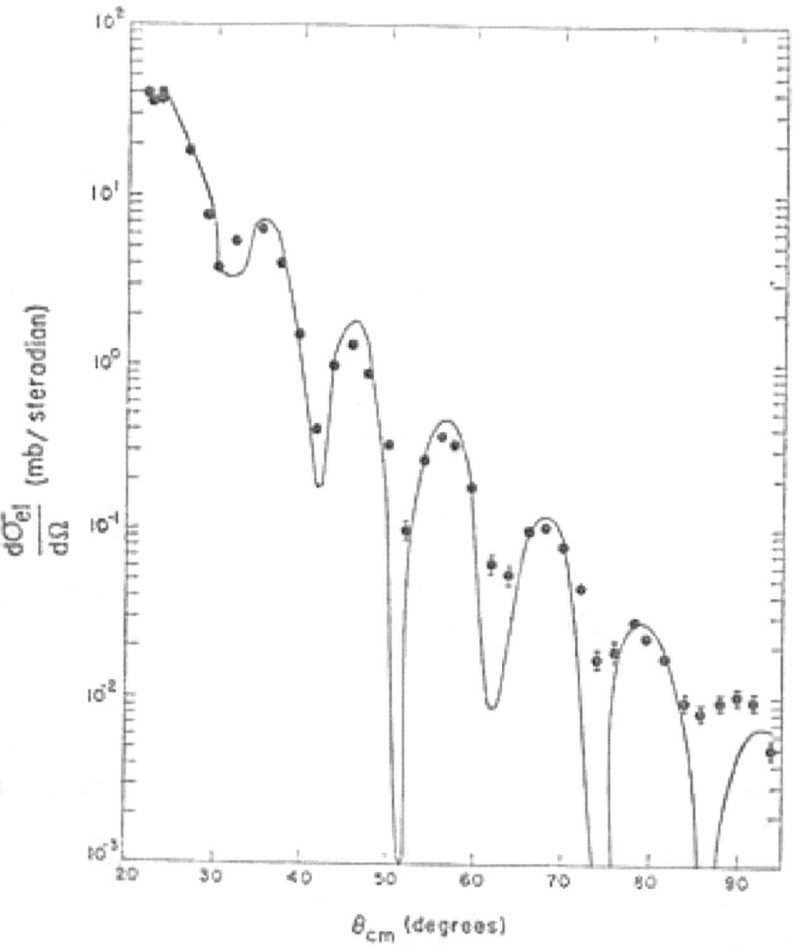

Figure 6.16d. Elastic scattering of α particles by Cu

A measurement of 40.2 MeV α particle by Cu as a function of scattering angle reported by Eisenberg and Porter (6.43) is shown in figure 6.16d. The data is fitted with the optical model theory. Similar oscillatory pattern of differential scattering cross section is observed in the inelastic scattering of α particles in the nuclei. This pattern depends upon the α particle energy; the higher the energy, the more pronounced is the oscillatory pattern.

The (α, p), (α, n) and (α,d) Reactions

The α particle-induced reactions such as (α, n) and (α, p) and (α, d) at energies above 10 MeV are usually direct type of reactions where the α particle enters the target nucleus and knocks either the neutron or proton from inside the nucleus, and it is captured by the target nucleus. Measurement of energies of the outgoing particle and the information of Q-value of the reaction can provide the information about the excited states of the residual nucleus.

The α particle is a very tightly bound nucleus; therefore, it will not be easily stripped of its constituents.

A typical result of (α, p) reaction of 30.4 MeV α particles on P^{31} leading to the ground state of the residual nucleus S^{34} is shown in figure 6.16c.

This shows that the differential cross section of protons emitted in the reaction as a function of angle in the center of mass displays a diffraction pattern similar to what is observed in α particle elastic and inelastic scattering. The fit to the data is made on the basis of direct reaction theory of Butler et al. (6.23) using optical model parameters.

Similar results are observed in the (α, n) and (α, d) reactions. (α,d) reaction on N^{14} and Li^6 was studied by Zeidman and Yntema (6.47) for incident α particle energy of 43 Mev. The angular distribution of deuterons in N^{14} (α,d) reaction exhibit oscillatory pattern of forward peaking with two additional secondary maxima at 40 and 60 degrees.

A fit to the angular distribution was made using spherical Bessel function under the assumption of a knock out process in which the α-particle enters the nucleus and knocks out a deuteron. Inclusion of small contribution from stripping reaction improves the fit to the data. In the (α,d) reaction on Li^6 the angular distribution can be also fitted with a mixture of knock out and α particle stripping.

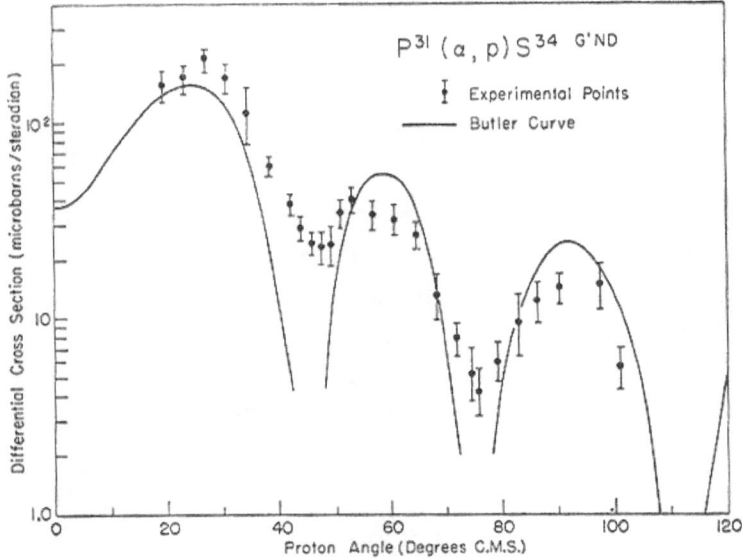

Figure 6.16 c. Differential cross section of P^{31} (α, p) S^{34} reaction to ground state of residual nucleus

6.33. Coulomb Excitation

One of the ways to study nuclear energy levels is to produce electromagnetic excitations of the nucleus by bombarding the target nucleus with high energy charged particles such as protons. This process is called Coulomb excitations. This involves excitation of the target nucleus from the ground state to excited states from which the nucleus decays to ground state by the emission of a gamma ray or cascade of gamma rays.

The incident charged particle loses some of its energy to the target nucleus but does not penetrate the nucleus involving no nuclear interaction.

These reactions have large cross sections in nuclei, which show features of collective model. The cross section increases linearly with energy of the incident particle. However, Coulomb excitations of energy levels in heavy nuclei can be executed by low energy charged particles since excited states in heavy nuclei have very low energies.

6.34. Photonuclear Reactions

This reaction is inverse of radiative capture of a particle by a nucleus.

In this nuclear reaction, a γ ray can interact with a nucleus, causing the emission of particles from inside the nucleus.

The simplest photodisintegration process is that of deuteron. Binding energy of deuteron is -2.22 MeV. A gamma ray of energy above the binding energy can break up the deuteron into its constituent particles—a neutron and a proton. The cross section for the photodisintegration of deuteron by γ rays emitted from radioactive nuclide taken by Evans (6.30) is shown in figure 6.17.

The cross section shows a broad resonance with maxima at $E_\gamma =$ 4.2 MeV. Points are experimental data, and solid curves are theoretical predictions based upon the following.

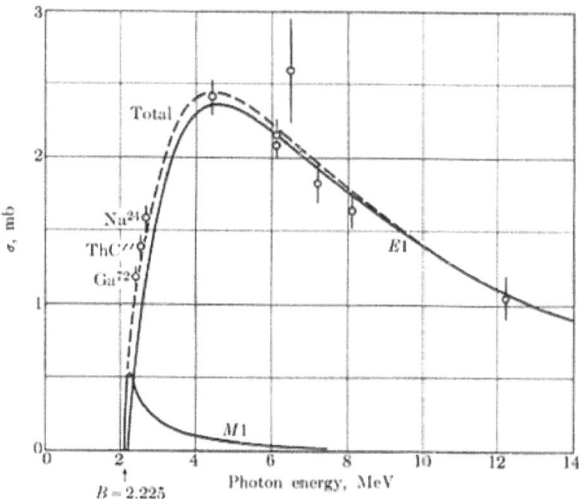

Figure 6.17 Cross section for the photodisintegration of the deuteron. The points represent experimental data; the three first were obtained by use of gamma rays from radioactive nuclides. (Reproduced by permission from R. D. Evans, *The Atomic Nucleus*, New York: McGraw-Hill Book Company, 1955.)

The reaction mechanism can be thought of as a transition from the ^3S ground state of deuteron to a ^1S virtual state (there is no excited state of deuteron) by the absorption of a magnetic dipole (M1) radiation or a transition from a ^3S state to a ^3P virtual state by the absorption of an electric dipole radiation. The magnetic dipole radiation (M1) represents $L = 1$ and change of parity. Hence, going from ^3S state to ^1S state requires $L = 1$ and no change in parity. A transition from ^{3}S to ^{3}P state will require angular momentum of L = 1 and change of parity, which is the characteristic of E1 gamma radiation.

Similar photonuclear reactions in other nuclei have been studied. One example is photonuclear reaction cross section in O_8^{16} taken by Bramblett et al. (6.31) up to $E\gamma$ = 30 MeV is shown in Figure 6.18.

This shows many resonances after monoenergetic gamma ray energy is varied between 16 and 30 MeV in small intervals. Resonances occur when the threshold energy for (γ, n), (γ, pn) and $(\gamma, 2n)$ reactions are reached as seen in the figure.

A broad resonance centered at about E_γ = 24 MeV is known as a giant dipole resonance, which is observed in many nuclei at an energy varying slowly with the nucleon number A. Scientists believe that the giant resonance occur due to the collective harmonic motion of protons relative to neutrons.

In the study of Pb^{208} (γ, n) Pb^{207} reaction near threshold energy, Toohey and Jackson (6.42) show emission of neutrons of different energies leading to the excited states of Pb^{207}. These results show that there exists a concentration of ground state M1 radiation strength in Pb ^{208} centered at an excitation energy of 7.9 MeV. The strength is spread over seven resonances in the energy range from 7.4 MeV to 8.25 MeV. The sum of radiation widths is about 50% of the Weisskopf limit.

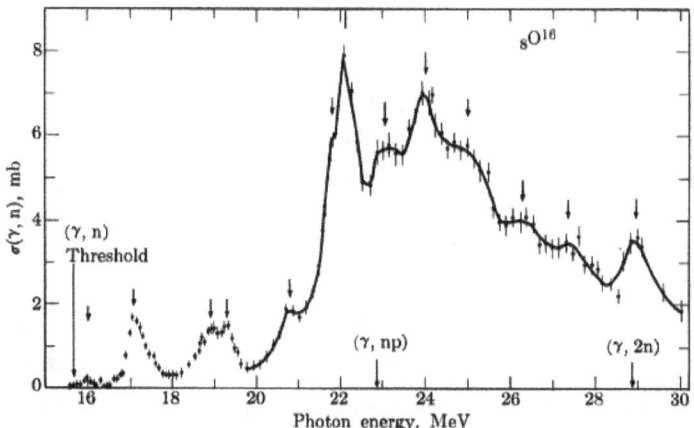

Figure 6.18 Photoneutron cross section of O^{16} up to 30 MeV, showing resonances from levels in O^{16}. [From R. L. Bramblett, J. T. Caldwell, R. R. Harvey, and S. C. Fultz, *Phys. Rev.* **133**, B869 (1964).]

6.35. Nuclear Fission

Another nuclear reaction that has played an important role in the world is that of the fission of nuclei. Some stable heavy nuclei, when they absorb neutrons or protons, undergo fission, meaning that the nucleus fragments into pieces with release of energy. In general, there are two kinds of fission processes. One is known as spontaneous fission, and the other is known as induced fission.

6.36. Induced Fission of Uranium with Neutrons

In 1939, during world war, scientists, Otto Hahn and F. Strassman (6.32), discovered that alkaline earth metals were produced when uranium nucleus was irradiated by stray neutrons. The explanation of this reaction was given by Lise Meitner and Otto Frisch (6.33) who theorized that the uranium nucleus after absorption of neutrons underwent fission, breaking up in two fragments of lower masses. He also noted that if a heavy nucleus breaks up into smaller nuclei, the total binding energy of uranium nucleus minus the combined binding energies of smaller fragments would be released as kinetic energy, which was observed in this reaction. Total energy released was about 220 MeV per nuclear fission of U^{235}. It was also noted that since heavy nucleus had more neutrons than the smaller mass nuclei, there would also be emitted few extra neutrons after the process of fission. A theoretical explanation of this reaction was given by Bohr and Wheeler (6.34) in terms of a liquid drop model discussed in chapter 4 section 4.20.

According to the model, which is based on hydrodynamic model of a drop of liquid, the terms leading to the fission of an excited spherical nucleus involves surface and Coulomb energy terms. For a spherical nucleus, the sum of these two terms can be written as

$$E = 4\pi R^2 S + \frac{3k}{5}\frac{(Ze)^2}{R} \qquad (6.40)$$

where S is surface tension and R is nuclear radius.

By using the constants in semiempirical mass formula discussed in section 4.20, Bohr and Wheeler (6.34) found that the ratio of strength of Coulomb force to the strength of surface tension force is Z^2/A. When this ratio is equal to 47.8, the spherical liquid drop becomes unstable and gets deformed. The nucleus becomes more and more deformed and reaches a saddle point from which it leads to fission, breaking apart in two or more fragments thereby releasing a large amount of energy.

The process of fission is not well understood. The evolution of a nucleus undergoing fission requires the knowledge of the nuclear states at the saddle and scission points. Information about the saddle point states can be obtained by analyzing the fission cross section and fission fragments angular distribution. The information about the nuclear states at the scission point can be determined from the properties of fission fragments such as their masses, their kinetic, and excitation energies. Many experiments have been performed, and the results indicate that the fission process changes from a super fluid behavior at low energy of excitation to a viscous behavior at higher excitation energies.

In order for the fission of the nuclei to occur, fission fragments have to overcome the Coulomb barriers to be emitted. It is only nuclei heavier than thorium that the energy released in the fission is large enough to overcome the Coulomb barrier.

An example of fission of U_{92}^{235} with low energy neutrons is given as follows:

$$U^{235} + n \rightarrow U^{236} \rightarrow X + Y + \gamma_n \qquad (6.41)$$

The capture of a neutron by U^{235} leads to the formation of the compound nucleus U^{236}, which breaks up in two large but unequal mass fragments X and Y. The γ_n is the number of neutrons emitted in the fission process. Distribution of masses of fragments lie in two groups: light and heavy nuclei. The sum of these two fragments is the same in all cases.

A typical distribution of masses of fission fragments as a function of atomic mass A induced by thermal neutrons in U^{235} is shown in figure 6.19.

This figure shows that the maximum probability of mass distribution occurs with two fragments—one fragment with a mass of $A = 140$ and the smaller fragment with a mass $A = 95$. It is not known as to why mass distribution of two fragments in fission induced by thermal neutrons is asymmetric whereas fission induced by higher energy neutrons as in the case of fission of U^{238} is observed to have symmetric distribution of masses.

In a typical fission of U^{235} induced by thermal neutrons, X and Y can be Rb^{93} and Cs^{141} nuclei. This division of masses releases about 180 MeV energy. Other distribution of masses yields roughly the same magnitude of energy up to a maximum of about 220 MeV.

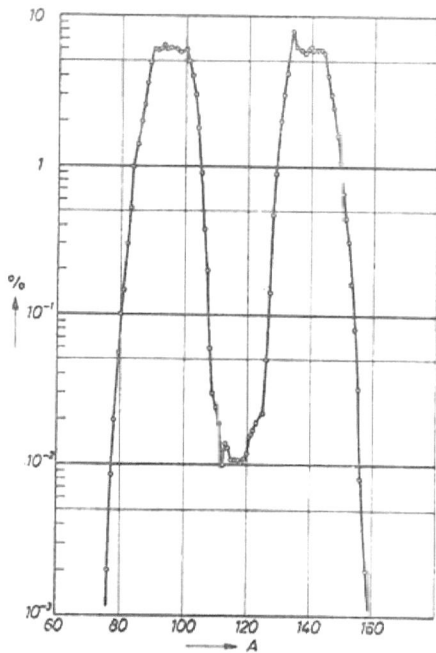

Figure 6.19. Distribution of masses of fragments in the fission of U^{235}

6.37. Energy Released During the Fission

Energy released during fission is divided among the two mass fragments: prompt neutrons and the recoiling nucleus. The two unequal fission fragments carry about 80% of the total energy in the form of their kinetic energies. The distribution of the kinetic energies carried by each mass fragment varies statistically for each fission.

Energy distribution of fragment masses in the fission of U^{235} is shown in figure 6.19(b). The average energy of the two fragments is about 61.4 MeV for the larger mass fragment and 93.1 MeV for the lighter fragment. Energy among the two unequal mass fragments is inversely proportional to their masses.

In addition, energy of about 5 MeV is carried by about 2.5 prompt neutrons emitted during the fission. The mean energy of each neutron is about 2 MeV per fission with an exponential distribution of energies.

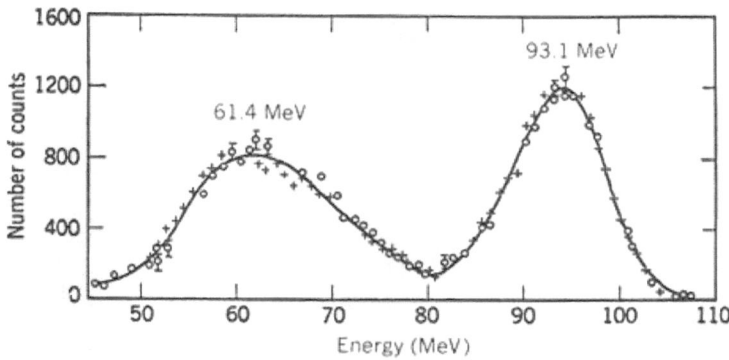

Figure 6.19 (b) shows the energy distribution among the two mass fragments. These measurements were made by Fowler and Rosen.

Fragment nuclei produced during the fission have more neutrons than stable nuclei. Therefore, they are highly unstable, and they can decay by successive emissions of β particles and γ rays. These nuclei have half-lives varying from few seconds to few years. Decay of these nuclei releases about 19 MeV in the form of kinetic energies of beta particles and accompanying neutrinos and about 7 MeV in the form of energies of γ rays.

These figures are only estimates, and the actual values depend on the fragment nuclei produced during the fission.

In some cases, if the β decay of a nucleus leaves the daughter nucleus in an excited state, which is above the separation energy of a neutron, a delayed neutron is emitted. It is estimated that about 0.65% of fission in U^{235} will emit delayed neutrons with half-lives varying from 0.23s to 55.7s.

6.38. Process of Neutron-induced Fission

Neutron-induced fission of transunic elements is rather a complex reaction. When a low energy neutron is captured by a heavy nucleus, a compound nucleus is formed. For example, the capture of a neutron by U^{235} forms the compound nucleus U^{236} at an excitation energy equal to the neutron binding energy. The ground state spin and parity of U^{235} is 7/2-. Hence, the capture of s-wave $(1 = 0)$ neutron will populate compound states either with J^π of 3⁻ or 4⁻ whereas capture of p-wave neutron will populate positive parity states. After the formation of the compound nucleus, its decay can take place via several channels such as elastic scattering of neutrons, capture, and fission. If the fission threshold is above the excitation energy of the compound nucleus, fission will not occur. However, if the fission threshold is below the excitation energy, the fission channel will open, and it will compete with other types of decay of the compound nucleus.

Scientists have determined that the effective fission threshold depends significantly on the spin and parity of the compound nucleus.

6.39. Statistical Model of Fission

The theory of fission to explain the asymmetrical nature of mass distribution of fragments was developed by Fong (6.35). The theory is based upon the decay of the compound nucleus formed by the capture of neutrons by uranium nucleus. The probability of the emission of fission products is considered by Fong to be proportional to the density of states at the excitation energy reached after capture. This determines the asymmetric distribution of fission fragments. Mass distribution curve calculated by Fong is in good agreement with the experimental measurements.

As the energy of excitation increases as in the case of the fission of U^{238} with fast neutrons of about 1 MeV, the mass distribution of fragments becomes symmetric.

The average number of neutrons emitted in fission is about 2.5 per fission, but this number varies slightly with the nucleus undergoing fission. The emission of more than one neutron in a fission promotes a chain reaction of fissions giving rise to uncontrolled production of energy, leading to an explosion. This discovery was made during the the Second World War, which culminated in the development of an atom bomb.

By controlling the number of neutrons, one can produce steady release of energy as is the case in an atomic reactor. The first atomic or nuclear reactor was built in 1942 by E. Fermi in Chicago.

Measurements of fission cross section in fissionable nuclei such as the isotopes of thorium, uranium, and plutonium have been extensively made by many scientists. These measurements show many resonances in each nucleus. The analysis of the cross section data using Breit-Wigner formula has provided important information about the average values of fission widths and strength function.

Fission cross section is a strong function of neutron energy. Figure 6.20 shows a plot of fission cross section measurements compiled by Hughes and Schwartz (6.36) as a function of neutron energy in U^{235} and U^{238}.

As seen in the figure, the fission cross section in U^{235} decreases rapidly by orders of magnitude as the neutron energy increases from 10^{-3} eV to about 1 eV. There are also many resonances observed at neutron energies between 1 eV and 100 eV. At higher neutron, energies resonances overlap due to a small level spaving of about 1 eV.

The fission cross section is negligible for U^{238} at low energies. The threshold of fission in U^{238} takes place at about 1 MeV as is seen in the figure. For this reason, the less abundant isotope of uranium with mass 235 and plutonium with mass 239 have large cross section of fission at very low neutron energies and are used for the generation of nuclear energy.

Many transuranic nuclei undergo spontaneous fission. Most of the nuclei beyond plutonium are unstable and fission spontaneously.

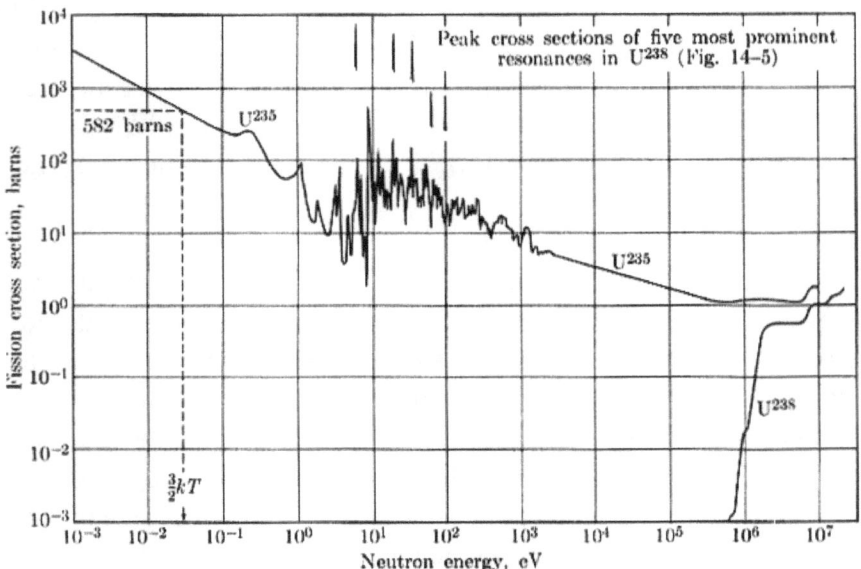

Figure 6.20. Fission cross section in U^{235} and U^{238} vs. neutron energy E_n. [Replotted from several charts in D. J. Hughes and R. B. Schwartz, *Neutron Cross Sections*, Brookhaven National Laboratory Report 325, 2nd. ed. (1958).]

High resolution measurements of neutron fission cross sections in U^{235} made by Cao et al. (6.45) are shown in figure 6.21.

This shows about 40 resonances in the energy interval of 5 eV and 40 eV with an average level spacing of about 1 eV in the compound nucleus U^{236}. These authors have measured the fission width of each resonance. The fission width varies widely from resonance to resonance. This is due to the statistical behavior of the decay of compound nucleus. The sum of these fission widths in a given energy interval provides values of the fission strength function, which can be compared with the prediction of the optical model similar to the case of neutron scattering strength function.

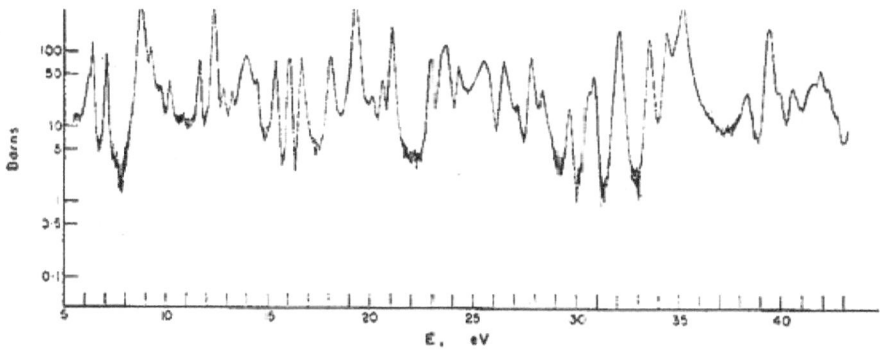

Figure 6.21. Fission neutron cross sections in U^{235} in the energy range of 5 eV to 40 eV

6.40. Fusion of Nuclei

One knows that the source of vast energy produced in our sun is the fusion of hydrogen, which our sun has in abundance. The fusion of two hydrogen nuclei requires energy to overcome the Coulomb repulsion between similarly charged particles. This energy can be provided in the laboratory by accelerating protons in an accelerator, or the energy can be provided by heat energy. According to the kinetic theory of gases, energy is related to the temperature by the relation $E = kT$, where T is the absolute temperature in Kelvin and k is Boltzman constant.

Therefore, one must supply kinetic energy to protons to overcome the Coulomb potential barrier. This energy is provided by ten million degrees of temperature existing in the interior of the sun.

Fusion of hydrogen to helium takes place in several steps of nuclear reaction as given below.

$$H_1^1 + H_1^1 \rightarrow H_1^2 + e^+ + \upsilon + 0.42 \; Mev \tag{6.43}$$

$$H_1^2 + H_1^1 \rightarrow He_2^3 + \gamma + 5.49 \; Mev \tag{6.44}$$

$$He_2^3 + He_2^3 \rightarrow He_2^4 + H_1^1 + H_1^1 + 12.86 \; Mev \tag{6.45}$$

Thus, eventually four hydrogen nuclei fuse together to produce He_2^4 and a total energy equal to 18.77 MeV as the sum of energies released in the above three reactions.

Measurement of p-p scattering cross section given in equation 6.43 has been extensively studied at many energies using particle accelerators. These studies have provided valuable information about the strength and range of strong nuclear force. This was discussed in chapter 3. However, the reaction 6.43 involves capture of two protons leading to a virtual state of diproton, which has never been observed. In this reaction, a proton converts into a neutron accompanied by a positron and a neutrino. Such a reaction is essentially β decay of a proton involving weak force. Decay of free protons has been investigated extensively and has never been observed. This reaction has very low cross section but takes place extensively in the sun and in billions of stars. This is due to extremely high density of protons present in these stars and extreme temperature of millions of degrees, which is needed to overcome the Coulomb barrier.

The positron emitted in this reaction annihilates with free electrons producing two 0.51MeV gamma rays.

Reaction given in equation 6.44 have been extensively studied at many proton energies using particle accelerators and targets of deuterium gas. This reaction forms He³, producing energetic γ rays.

Reaction 6.45 of the fusion of two He^3 nuclei have been studied in the laboratories at many energies.

6.41. Fusion Reactors

Fusion reaction has been considered as a source of clean power using vast resources of hydrogen present in water. However, for reasons given above, the energy required to overcome the Coulomb repulsion is so great that the temperature in excess of million degrees would vaporize all matter.

Such fusion of hydrogen was used in the development of a hydrogen bomb where the energy needed was provided by detonating a nuclear fission bomb with uranium nuclei.

Scientists have carried out extensive studies of developing electromagnetic devices to produce high density and high temperature plasma in the laboratories. Such studies have provided information about the reaction mechanism and cross sections; however, to use these devices in a practical application of producing fusion energy is not successful.

Fusion of heavy nuclei such as two He_2^4 nuclei requires much greater kinetic energy to overcome larger Coulomb barrier. In the interior of the sun at $T = 10^8$ K, such fusion of the nuclei takes place forming C_6^{12} nucleus. Fusion of heavier nuclei requires still higher temperatures, which is produced in high mass stars in the universe.

In the interior of high mass stars, fusion of C^{12} with He^4 can produce O^{16}, and fusion with He^4 can produce Ne^{20}, and so on all the way up to $Fe^{56.}$

When such stars die, these heavy nuclei are dispersed in the interstellar medium and then are captured by other terrestrial bodies like our planet.

6.42. Heavy Ion Reactions

Particles heavier than protons, neutrons, and deuterons are nuclei such as Li^7, Be^9, C^{12}, etc. These nuclei are accelerated as singly charged or multiple charged ions in an accelerator and can strike certain light and heavy targets.

These ions, in view of their high charges, have much higher Coulomb potential barrier to overcome before they can interact with the target nuclei.

Figure 6.22 shows for C^{12}, N^{14}, O^{16}, and Ne^{20} ions the energy in MeV to overcome the Coulomb barrier as a function of atomic number Z of target nuclei. Even for the lightest C^{12} ion and a low mass target $Z = 10$, minimum energy needed is about 20 MeV. These energies can only be available in high energy cyclotrons.

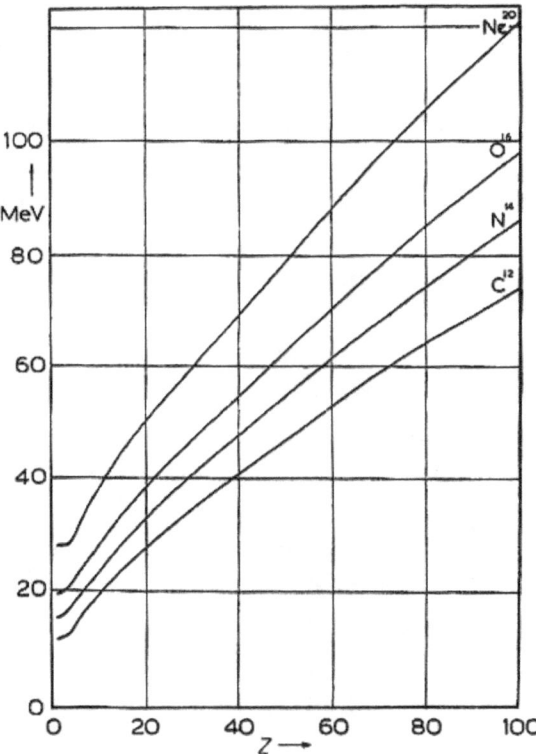

Figure 6.22. Variation with atomic number of target of the laboratory energy required for barrier penetration by heavy ions.

The nuclear reaction produced by the capture of these ions are manifold including fusion or capture to produce heavier nuclei as in the stars. These heavy ions striking a target nucleus can produce knockout or stripping reactions. Such heavy ion reactions can produce new nuclei such as neutron deficient nuclei, which are unstable and decay by α and β particles. Heavy ion reactions at high energies will produce spallation of the nucleus with the emission of many particles and the creation of new nuclei. The study of such nuclei will be of great interest to the nuclear physicists.

The excitation energy of the compound nucleus formed by the capture of these heavy ions by a nucleus will be very high and will create very high temperatures causing evaporation of many particles from inside the nucleus. The study of these types of reaction at high energy is interesting but very complex.

6.43. Summary of Nuclear Reactions

Study of nuclear reactions induced by neutrons and charged particles such as protons, deuterons, tritons, He^3 and He^4, and heavier ions in the energy range of thermal neutrons up to energies of 20 MeV have provided great insight into the mechanism of interaction of these nucleons with nuclei. These have yielded valuable information about the nuclear force and nuclear structure.

In general interaction of a single nucleon with a target nucleus produces capture of this nucleon by the target nucleus forming a compound nucleus. The resulting reaction is usually elastic scattering of the incident nucleon. If the kinetic enrgy of the incident particle is above the the the energy of the excited state of the target nucleusthe reaction would be inelastic scattering of the incident particle.

In a book on introduction to such topics, it is not necessary to indulge in too much theoretical discussion of each type of nuclear reaction. In this book, I have provided conceptual information as to how different nucleons interact with nuclei and what useful information can be obtained about the nuclear structure and nuclear decays.

I do not believe that it would be beneficial for students to be burdened with the discussion of such advanced topics unless the students have an interest in pursuing research in this field. There are books written on this topic exclusively such as the book on nuclear reaction by Endt and Demeur.

CHAPTER 7

NUCLEAR PHYSICS APPLICATION

Knowledge of nuclear physics has been applied to many fields, which provide great benefits to society. Some of these applications are in the field of medicine, environment, energy, and industry. Discovery of electron and photoelectric effect allowed the development of cathode ray tubes, which helped to develop oscilloscopes and television screens. Development of nuclear magnetic resonance in liquids and solids helped to develop the instrument MRI for the diagnosis of diseases. Discovery of nuclear fission in uranium has led to the development of nuclear energy.

7.1. Biomedical Applications

In order to detect the diseases that attack human beings, it is necessary to take images of our bones, our organs, and the working of our bodily functions. The images of a given object is produced either by reflection of radiation or by the detection of the source of radiation. Ordinary light cannot penetrate the skin, and hence, it takes only picture of the exterior of the human body. However, x-rays with much greater energy can penetrate the skin and the flesh, but it cannot penetrate through the bones. Hence, these x-rays are reflected by the bones allowing one to take images of the bones. This application is useful for determining bone fractures or bone deformities.

7.2   CT or CAT Scanning

The improvement in the use of x-rays as a diagnostic tool is the CT scan known as computerized tomography or the advanced CAT scan known as computerized axial tomography. In this technique, beams of x-rays are directed onto the patient from different directions. The device consists of several x-ray tubes located inside a ring through which the patient is moved. Each of the x-ray beam is detected by a film located at the other end. Intensities of beams detected are different, depending upon the nature of the body material. The complete images obtained are then fed to a computer, which can provide a three-dimensional image of the human body. This device is now commonly used for diagnosis of the diseases of internal organs of human body.

7.3. Use of Radioactive Isotopes

It was after the discovery of radioactivity that one was able to use high energy gamma rays in medical field as a diagnostic tool. Gamma rays in view of their greater energy can penetrate through bones and can provide images of the organs and the entire body.

7.4. Positron Emission Topography

This technique provides an important medical diagnostic tool. Certain isotopes such as $_8O^{15}$, $_7N^{13}$, $_6C^{11}$, etc., decay by emitting positrons. All these nuclei that have short half-life in minutes can be produced by nuclear reactions of charged particles with suitable nuclei. The charged particles such as protons, deuterons can be accelerated to about few MeV energies in a Van de Graaff generator or a low energy cyclotron. These isotopes are injected into the body where they are deposited in specific locations. Emission of positrons during β decay encounters an electron in the body tissue. Electron and positron pair then annihilates and produces two 0.51 MeV γ rays. These are emitted in opposite directions. These gamma rays are then detected by a gamma ray detector placed in opposite directions. One can thus determine the line in which the annihilation occurs. Such information leads to an image developed by a computer, which points the location of the deposit of the radiation source.

7.5. NMRI (Nuclear Magnetic Resonance Imaging)

The nuclear magnetic resonance technique has been used for the measurement of magnetic moments of nuclei. This technique is now being used in medical diagnosis. The basic principle on which the technique is based is the property of protons, which are the nucleus of the hydrogen atoms. Our body contains billions of hydrogen atoms. The proton can exist in two separate spin states: spin parallel and spin antiparallel to the magnetic field. The energy of the parallel orientation is $+\mu_p B$ and that of the antiparallel state is $-\mu_p B$, where μ_p is the magnetic dipole moment of the proton and B is the applied magnetic field. This energy difference between the two orientations is then $E = 2\mu_p B$. If an RF signal of a frequency $\upsilon = 2\mu_p B/\hbar$ is applied, the protons will absorb this energy as they flip from one orientation to the other. This frequency is known as resonant frequency.

The nuclear magnetic resonance phenomenon is utilized in developing an instrument known as MRI to take images of the human body. In general, human tissues that has the least number of hydrogen atoms such as bones appears dark and tissues that has many hydrogen atoms such as fatty tissues appears very bright in the images.

The patient lies inside a large cylindrical-shaped magnet. Radio waves are passed through the body. This affects the body's hydrogen atoms forcing their protons into a different position. As these nuclei move back into place, these generate radio signals, which are detected and fed to a computer. The computer generates them into a very clear image of the human body. The images depend on the location and strength of the generated signals.

7.6. Treatment of Tumors or Cancerous Growths

The main use of gamma rays produced by radioactive nuclei has been in the treatment of tumors. When energetic radiation interacts with our cells, it imparts energy to cells thereby raising their temperature. This causes the cells to produce free radicals to die. Due to the availability of particle accelerators and nuclear reactions, scientists produced many unstable nuclei producing radioactive decays. In general, these nuclei decay by beta rays followed by the emission of gamma rays. These unstable nuclei have different lifetimes varying from few seconds to several years. Physicians have been using these radioactive sources for the treatment of cancers. One of the drawbacks of this application is that such radiation, in addition to damaging the cancerous cells, can also damage the healthy neighboring cells with serious consequences. An external source of gamma rays produces radiation in all directions; hence, it is necessary to provide collimator to provide as narrow beam of radiation as is possible and to direct the radiation to the tumors with a pinpoint precision. This is known as shaped beam radiation therapy. The treatment involves laying the patient on a robotic couch where it automatically aligns the radiation beam with the tumors. The system continuously shapes the beam to match the shape and size of the tumors from all angles. The surrounding normal tissues and delicate structures such as brain stem, spinal chord, and optic nerves are spared from damage. It is an outpatient painless procedure, which is offered only in some hospitals.

7.7. Other Radiation Treatments

The human body contains about 80% water, which is a molecule of hydrogen and oxygen. In addition, the human body contains trace amounts of other essential elements such as Li, C, N, Na, Mg, P, S, Cl, K, Ca, Fe, Cr, Ni, etc.

Deficiency or excess of any of these elements in our body can cause sickness and diseases. It is therefore necessary to map the concentration of such elements. Neutron and charged particle activation techniques are very useful in determining these deficiencies.

One other alternative is to inject the radioactive substance in the bloodstream. The radioactive substance could be chosen so that it has a tendency to accumulate in one organ and not be distributed everywhere. The radioisotopes, which are commonly used, are K^{47}, Co^{57}, Co^{60}, Fe^{59}, Ga^{68}, I^{131}, Ir^{192}, and Pd^{03}, etc.

The recent development in treating cancers is the use of particle accelerators, which provides high energy charged particles and heavy ions. Such particles from an accelerator can be collimated to a very fine beam using magnets. Interaction of charged particles such as protons, deuterons, α particles, and heavy ions with matter produces ionization and nuclear reactions such as (p, α), (p, n) and (α, n), and γ rays characteristic of the element producing such radiation. A detector can detect these γ rays and thus identifies the sample. These particles can be directed to the exact location of the tumor in a pinpointed manner. The treatment with a high energy proton beam is more extensively used because the procedure produces less damage to the healthy tissues while killing the cancerous tissues.

7.8. Nuclear Energy

Nuclear energy is produced by fission of uranium or other transuranic nuclei. Element uranium is found in our planet in large abundance; however, it has two isotopes $_{92}U^{235}$ and $_{92}U^{238}$ with abundance of 0.72 and 99.28%. Scientists discovered that U^{235} has a very large fission cross section at essentially thermal energy of neutrons whereas U^{238} has insignificant fission cross section for low energy neutrons. The threshold for fission in U^{238} is at the neutron energy of about 1.4 MeV. When the nucleus of U^{235} captures a low energy neutron, it forms a compound nucleus with an excitation energy of 6.46 MeV whereas the capture of a neutron by U^{238} excites the compound nucleus to the energy of 4.78 MeV. This energy difference requires high energy neutron to produce fission in U^{238}. However, this isotope has a large resonance capture cross section. The capture of neutrons by U^{238} produces Pu^{239}, which has a large fission cross section at thermal neutron energy. In general, even-odd nuclei undergo thermal neutron fission whereas even-even nuclei do not.

As discussed earlier in section 6.35, the fission of uranium nucleus breaks up in two nearly equal mass nuclei and emits on the average about 2.5 neutrons for each fission. The use of fission for producing nuclear energy requires good control of these emitted neutron population. The usual requirement for nuclear energy production in a controlled manner is that there is one neutron per fission. Thus, one needs a mechanism to absorb extra neutrons to produce the 1:1 ratio. Second aspect of fission reaction is that emitted neutrons have high energies and thus these neutrons have to be slowed down to low energies for which the fission cross section is very large. These requirements are met in a nuclear reactor as was shown by Enrico Fermi in 1942.

Over the years, nuclear engineers have developed many types of reactors—some are experimental for research purposes and some are built to produce nuclear power. Some of the common types of reactors are given below.

7.9. Boiling Water or Pressurized Water Reactors

These are the earlier types of power reactors. Basically, the fuel is enriched U^{235} to about 3% of mass. The core of the reactor consists of a number of fuel rods containing pellets of uranium oxide encashed in metal tubes. These rods are suspended in a tank filled with water. To control the neutron number and the rate of fission, cadmium, which has a large neutron capture cross section, are used as control rods. These rods can be inserted to reduce the number of neutrons or pulled to increase the number essentially controlling the rate of nuclear fission.

In order to reduce the neutron energy to thermal energy, water is used as a moderator and also as cooling substance. We know that fission releases a lot of energy heating the fuel rods. Hence, it is essential to cool the system to avoid fuel meltdown, which will cause a serious accident and release hazardous radioactivity to the environment. It was realized that water used as moderator has a serious drawback that the hydrogen in water captures neutrons to form deuterium with the release of 2.2 MeV gamma rays. This will cause loss of neutrons.

7.10. Heavy Water Reactors

In order to avoid the problem of neutron capture by hydrogen, a new design of reactors was developed. This reactor instead used heavy water with D_2O as moderator and cooling substance. Deuterium has a very small neutron thermal capture cross section. Reactors using heavy water were developed in Canada known as CANDU reactors. This type of reactor uses natural uranium as fuel and pressurized heavy water as moderator and cooling substance. Cadmium rods were used as control rods.

Further improvement was made in cooling design of reactors by using pressurized steam or gases. These were found to be very successful.

Since natural uranium contains 99.28% of U^{238}, scientists explored the possibility of using natural uranium to produce nuclear power. The capture of neutrons by U^{238} produces Pu^{239} in two successive β decays accompanied by antineutrinos υ as shown below.

$$_{92}U^{238} + n = {}_{92}U^{239} + \gamma \text{ rays}$$
$$_{92}U^{239} = {}_{93}Np^{239} + e^- + \upsilon$$
$$_{93}Np^{239} = {}_{94}Pu^{239} + e^- + \upsilon$$

where $_{94}Pu^{239}$ is a fissile element. The reactor design that emphasizes the production of Pu^{239} are used as new types of reactors. These are as follows:

7.11. Liquid Metal Fast Breeder Reactors

The primary fissile nuclide for these types of reactors is Pu^{239} and fertile nuclide is U^{238}. The reactor uses as fuel 30% of Pu^{239} and 70% of U^{238} as pellets encased in stainless metal tubes. Liquid sodium is used as neutron moderator as well as cooling substance. Reactivity control is achieved by control rods made up of B_4C. In this type of reactor, U^{238} after neutron capture is converted to Pu^{239}, which provides the energy by thermal neutron fission as discussed in section 7.10. The development of fast breeder reactors is very important in view of the utilization of all natural uranium found on our planet for the production of nuclear energy.

Our planet has abundant supply of element Th^{232}, which has a low cross section for fission at low-neutron energy. The capture of a neutron by this nucleus produces U^{233}, which has a large fission cross section at thermal neutron energy. Hence, a similar type of reactor can be used with a mixture of U^{233} and Th^{232} as fuel to produce nuclear energy. The conversion takes place as follows:

$$_{90}Th^{232} + n = {}_{90}Th^{233} + \gamma \text{ rays}$$
$$_{90}Th^{233} = {}_{91}Pa^{233} + e{-} + \upsilon$$
$$_{91}Pa^{233} = {}_{92}U^{233} + e{-} + \upsilon$$

7.12. Safety of Reactors

The public has great concern for the development of nuclear reactors and hence, since 1970, no new reactors were approved by the government. It has now been realized that our country cannot depend upon foreign sources of energy, and our president is interested in renewing the development of nuclear energy.

The great concern for the safety of reactors is the accident similar to three-mile reactor and the Chernobyl reactor in Soviet Union. The other concern is the safe disposal of radioactive material produced inside the reactors. Fission of uranium produces nuclei, which are radioactive with various half-lives, some shortlived and some have very long life. The release of such radioactivity in the environment can cause catastrophic consequences for mankind. Therefore, it is absolutely essential that scientists find adequate means of disposing this material in a safe manner. It appears that we have acquired adequate knowledge of accomplishing this task.

7.13. Radioactive Dating

Radioactive dating is used in determining the ages of archeological and geological samples. The technique is based upon the determination of abundance of certain organic samples as a function of time. All radioactive nuclei decay with a certain half-life.

The most common radioactive nucleus used for this purpose is C^{14}, which has a half-life of 5,730 years. In natural carbon, this isotope is present in the amount of one atom for every 8.3×10^{11} atoms of C^{12}. This amount of C^{14} was present in all living organism at the beginning of creation of life. After the death of a living organism, this radioactive isotope decays with its half-life. The determination of the abundance of this isotope in the dead organism with the knowledge of its half-life can determine the age of that organism.

7.14. Radon Gas in Homes

Radioactive radon gas $_{86}Rn^{222}$, which has a half-life of 3.83 days is produced from the α particle decay of $_{88}Ra^{226}$, which is a by-product of the decay of $_{92}U^{238}$ by successive α decays. Uranium has a half-life of about

5 billion years and is found in abundance in the earth's crust. It is then transferred to the soil and the cement of homes. Since radon is a gas, it can easily mix with air and smoke and can lodge in our lungs. This can then damage our tissues.

Presence of this radioactive gas in our homes can be detected by using a γ ray monitor.

7.15. Application of Nuclear Physics in the Field of Environment and Crime Detection

In the last few decades, scientists have used particle accelerators and nuclear reactors for the study of pollution in our environment. The large emission of carbon oxides, sulphur, lead, and other hazardous elements from exhaust of gases by all kinds of vehicles and from coal-fired power plants has been steadily increasing and causing global warming. The increased concentration of these pollutants in our air, water, and food supplies has serious consequences for the quality of life and even survival of the mankind.

In order to monitor the growth of these pollutants, scientists have used nuclear techniques. The most used technique is the neutron-activation analysis (NAA). In this method, the samples of air, food, and water are irradiated by thermal neutrons from a high flux neutron source mostly from nuclear reactors. When an element captures neutrons, it is excited to a higher energy levels. After a very short time, this nucleus decays in an exponential manner by emitting characteristic γ rays. These gamma rays are detected by a NaI (Tl) scintillation detector, and measurement of their characteristic energy identifies the element present in the sample. Since the reactor provides a large flux $10^{12}-10^{14}$ neutrons, even a very small sample receives enough activation to be tested. From the measurement of the counting rate and the half-life of decay, one can determine the amount of the element present in the sample.

This technique has been used for decades to analyze the presence of trace elements in our environment and in solving crime by determining the presence of arsenic and mercury in the human hair and fingernails, etc.

Scientists are also using particle accelerators to produce beams of charged particles such as protons, deuterons, α particles, and fast neutrons from nuclear reactions. When a given element is bombarded by these particles, inner shell electrons are emitted, which produce ionization as well as x-rays characteristic of the element present in the sample. From measurement of energies of x-rays, one can determine the presence of element.

This technique known as PIXE has also been used in identifying certain elements in our blood.

7.16. Industrial Application of Nuclear Physics

Use of accelerators in the modification and quality control of materials is well established.

Ion implantation is used to modify the strength of steels and other structural materials. Some of the modified alloys are used in artificial joints, hips, and knees of human body.

A technique of combining coating and ion implantation is used to improve surface properties of a wide variety of materials.

A regular use of radiation in food products is well known.

7.17. Application of nuclear physics in astrophysics

It is well established that our universe was created about 14 billion years ago by the explosion of tremendous amount of energy in a so called "Big bang". The energy converted into matter dictated by the famous Einstein formula $E = M c^2$. At its creation the universe was very hot with temperature of about 10^{32} Kelvin degrees. First the energy created particles responsible for the gravity, electro-magnetic, strong and weak nuclear forces. This was followed by the creation of nuclei, atoms and finally matter in the form of huge mass of hydrogen atoms. This mass gathered together due to attractive force of gravity. The universe kept on expanding and cooling to very low temperatures. Large mass of hydrogen formed clouds which collapsed under force of gravity generating tremendous amount of heat energy and raising the temperatures of accretion disks to millions of degrees. This gave rise to the formation of stars, planets, moons, asteroids, comets and meteorites etc. in the universe. The knowledge gained in nuclear physics has helped to understand these phenomena.

Evolution of low and high mass stars from their creation to their demise have been studied in great length by astro -physicists. Now we know that the evolution of these stars follow different and complex paths. Study of the formation and demise of billions of stars in our universe requires knowledge of nuclear reactions. It is known that stars are formed when hydrogen nuclei fuse together and their fusion creates about 20Mev energy. However in view of their strong coulomb repulsion fusion of protons requires very high temperatures of about 10 million degrees Kelvin to bring the protons close together to fuse. The nuclear reactions of hydrogen nuclei to produce helium nucleus is described in section 6.38 of this book. These nuclear reactions produce positrons, neutrinos and gamma rays. The reaction leads to a chain reaction with continuous release of energy for many billions of years. It is also known that high mass stars can produce fussion of heavier nuclei such as helium, carbon, oxygen all the way up to iron. Beyond iron the nuclear reactions are carried by capture of protons, neutrons and α- particles. Demise of high mass stars involve photo disintegration of iron nucleus with the release of huge flux of positrons, neutrinos, antineutrinos and γ- rays.

Extensive studies of all these reactions have been carried out by nuclear scientists using particle accelerators and the knowledge gained from these studies is able to explain the processes taking place in stars. The final death of stars follow a process which involves many nuclear reactions ending in a violent explosion of the stars as nova or supernova. Some of these final stages of stars result in the formation of neutron stars and black holes whose physics has been studied in great detail. It is called a neutron star because the density is close to the density of a neutron. Black hole is a singularity with the gravity force such that even the mass less photons are not allowed to escape.

Scientists have discovered that our universe contains much more mass that can be observed by known techniques. It is now believed that about 90 % of mass of the universe is dark matter that is invisible. The nature of this matter is unknown but one believes that most of this matter may be non interacting neutrinos and antineutrinos. Neutrinos were discovered in decays of beta particles and in the decay of muons. Such studies are of great interest in nuclear physics.

Summary

Detailed discussion of these applications of nuclear physics to many other areas described in this chapter is beyond the scope of this book. I have given only an introduction to these areas. The interested readers should consult specialized books on these topics to gain more knowledge in these areas.

Closing Remarks

This book deals with the main and important topics of nuclear physics, which would provide a good background for students pursuing a higher degree in physics. Students who wish to pursue research in nuclear physics will need to take advanced courses in theoretical and experimental nuclear physics. It is however important to note that irrespective of the field in physics which a student may pursue for a Ph D. degree he or she must learn about the fundamental concepts of nuclear physics. With this in mind I have written this book to cover the basic concepts of nuclear physics in a one semester course of four months duration.

Even though in recent several decades many advances have taken place in high energy nuclear physics and also in heavy ion physics. I do not believe that it would be beneficial for students to be burdened with the discussion of such advanced topics unless the students have interest in pursuing research in this field. Both fields of particle and nuclear physics are highly developed and sophisticated. I do not find myself sufficiently qualified to teach a graduate level combined course of elementary particle and nuclear physics. One can however teach a general education course combining the two separate fields. I had given such a course "physics for humanists" at my university for several years for non science majors. For this course I used an excellent book "From quarks to cosmos" authored by Leon Lederman.

CHAPTER 8

Suggested Problems in Each Chapter Based Upon the Concepts of Nuclear Physics Discussed in This Book

8.1. CHAPTER 1

Q.1. Taking O^{16} nucleus as an example, describe the following properties of this nucleus.
  a) Nomenclature
  b) Size and shape
  c) Binding energy per nucleon (use atomic mass tables)
  d) Angular momentum
  e) Nuclear energy level

Q.2. Following nuclei are given.
  $_5X^{11}$, $_{16}X^{32}$, $_{29}X^{63}$, and $_{94}X^{239}$
  a) Calculate the number of protons and number of neutrons in each nucleus, and give the names of these nuclei.
  b) Using atomic mass tables, calculate the binding energy per nucleon of each nucleus.
  c) Calculate the radius of each nucleus assuming $r_0 = 1.30$ fm.
  d) Calculate the surface area of each nucleus.

Q.3. Hofstadter and his group made measurements of nuclear size.
  a) Describe their experiment.
  b) Discuss what information did they obtain regarding their sizes.
  c) How does the nuclear radius depend upon atomic mass?
  d) Describe also other methods of measuring nuclear size and how these measurements differ with Hofstadter's measurement.

Q.4. Using C^{12} nucleus, discuss the following:
a) All the basic properties of the nucleus.
b) If atomic mass of the nucleus C^{12} is 12.000u, $m_n = 1.008665u$ and $m_p = 1.007277u$, calculate the binding energy per nucleon of this nucleus.
c) Draw a diagram showing BE/nucleon of all nuclei as a function of atomic mass A.

Q.5. Do the following activity:
a) Discuss the concept of parity for nuclear wave functions.
b) If there are two particles in a nucleus, what would be the parity of the nucleus?
c) Is parity conserved for all nuclear reactions and decays? Explain any violation of this law.
d) Discuss the symmetry property of wave functions of two identical particles.

Q.6. Answer the following:
a) Describe the experiment that Lord Rutherford and his colleagues carried out to determine the nature of the nucleus.
b) On what considerations they concluded that the nucleus was much smaller in size than the atom?

Q.7. Describe briefly the different types of nuclear reactions that can take place when a nucleon such as a proton, neutron, or γ rays strike a target nucleus. Give an example for each case.

Q. 8. Nucleus consists of neutrons and protons. Motion of these particles produce currents and give rise to magnetic moments. Describe how one can calculate the magnetic moments of nuclei.

Q.9. Some nuclei are stable, and some are unstable.
a) State the conditions under which a nucleus will become unstable.
b) Describe various modes of decay of an unstable nucleus by giving examples for each type of decay.

Q.10. Answer the following:

  a)   What are the isotopes? Give an example.

  b)   What nuclei are isobars? Give an example.

  c)   What are mirror nuclei? Give an example.

  d)   The excited states of mirror nuclei have small energy difference. What is the reason for the difference?

Q.11. The energy for β transition between mirror nuclei N^{13} and C^{13} is 1.18 MeV. Determine the value of radius parameter r_0.

8.2. CHAPTER 2

Q.1. Deuteron is a composite nucleus with one proton and one neutron.

  a)   Describe the properties of deuteron such as mass, charge, size, binding energy, magnetic, and electric moments.

  b)   Discuss the experiments that will allow you to determine the binding energy of the deuteron.

Q.2

  a)   Describe how one can determine the nature of nuclear force from the properties of deuteron. These are binding energy, angular momentum, and its mean square radius.

  b)   Using Scrodinger equation of n-p interaction, derive a relationship between nuclear potential (V_0) and range (b) of the nuclear force.

Q. 3

  a)   Describe the nature of nucleon-nucleon force as it was proposed by Yukawa. Explain with the help of a diagram how this force originates between a neutron and a proton.

  b)   Discuss how Yukawa determined the range of this force from his theory.

  c)   Discuss the modern ideas about the nature of nucleons and the nature of the nucleon-nucleon force.

Q.4. The binding energy and the angular momentum ($J = 1$) is consistent with the assumption that deuteron is in a $^3S_1$ state; however, its magnetic movement is different from the sum of magnetic moment of a proton and a neutron. How do you explain the discrepancy?

Q.5. If the nuclear potential is $V_0 = 50$ MeV,
  a) calculate the wave number K associated with a nucleon of 5 MeV energy,
  b) calculate the wave number k associated with a particle of 5 MeV energy.

8.3. CHAPTER 3

Q.1. Nucleon-nucleon force can be determined by measurements of (n-p), (n-n), and (p-p) scattering cross sections.

  a) Describe briefly how these experiments are done.
  b) How do these experiments prove that nuclear force is charge independent?

Q. 2.
  a) What do you understand from the meaning of scattering length (a)?
  b) Show how you can derive a relation between phase shift δ and scattering length (a).
  c) What do you know about the sign of the scattering length? Draw a diagram to show how one determines the sign.

Q.3. The measured (n-p) scattering cross section at low energies is 20 barns. Whereas the theoretical value based upon the scattering length obtained from deuteron is about 4.0 barns.
  a) How do you explain this discrepancy?

Q. 4. In nucleon-nucleon scattering, cross section is related to phase shift δ_0 of the scattered particle given as $\sigma = 4 \pi \sin^2 \delta_0 / k^2$
  a) Describe the relationship of δ_0 and scattering length (a).
  b) What information about (a) is obtained from the study of (n-p), (n-n), and (p-p) measurements?

Q.5. Calculate the triplet (a_t) n-p scattering cross section at $E_n = 1.5$ MeV (lab) by using equations 3.17 and 3.18 for the square-well potential $V_0 = 73$ MeV and $b = 1.40$ fm and $c = 0.4$ fm.

Q.6 In (n-p) scattering, neutrons of incident energy of 4.4 MeV are scattered by the nuclear potential. If the value of this potential is 42 MeV, calculate the values of wave numbers k and K (use equation 2.11 and 2.13).

Q.7
 a)   How do you measure (n, n) scattering cross section?
 b)   What information about the scattering length and range of nuclear force is obtained from such measurements?

8.4. CHAPTER 4

Q.1
 a)   Describe the essential features of shell model of the nucleus.
      b) Describe the experimental data, which supports the shell and the data, which does not support this model. Give reasons for your answer.
 c)   Using $N = 1$ as the first shell, how many neutrons and protons can be accommodated in $N = 3$ shell? Explain.

Q. 2 Using shell model, determine $^\pi$ values of the ground states of nuclei: $_2He^3$, B^{16}, Na^{23}, Ca^{40}, and $_{20}Ca^{41}$, give the configurations of protons and neutrons for each nucleus.

Q.3
 a)   Write the radial wave function for $l = 0$ and $l = 1$ as given by equation 4.3.
 b)   Find the value of kr, which gives the first zero of the two functions.
 c)   Using the results of (a) and (b), calculate the distance in MeV between the lowest s-state and the lowest p-state in an infinitely deep well radius of 5.0 fm.

Q.4

  a)   What experimental data points to the fact that nuclei in rare
       earth region are permanently deformed? Discuss the type of
       deformations and what it means.
  b)   Give J^π values and energies of excited states of such even-even
       permanently deformed nuclei.

Q.5. Some medium weight nuclei show J^π of excited states as different from
shell model states even though these nuclei are not permanently deformed.
    Discuss the properties of such nuclei, and what are the excited states
of even-even nuclei?

Q.6. For odd-mass deformed nuclei, a unified model was proposed by
Nillson also known as Nillson's model.
  a)   Write the Hamiltonian for the model explaining each term.
  b)   What quantum numbers are conserved in such deformed nuclei?
       Show a diagram of a deformed nucleus with symmetery axis and
       quantum numbers.
  c)   For an odd nucleus in $d_{s/2}$ orbit, how will the energies in such an
       orbit be given in terms of quantum numbers Ω?
  d)   Supposing $_9F^9$ is such a deformed nucleus, what would its ground
       state J^π be?

Q. 7. Even-even permanently deformed nucleus Y^{166} has the following
excited state energies in (MeV) and J^π values. Calculate the energies in
MeV and J^π values of states with (?) marks.

E (MeV)	J^π
?	8+
?	?
0.330	4+
0.102	2+
0	0+

Q.8. By tabulating the possible m states of three quadrupole phonons ($\lambda = 2$) and their symemetrized combinations, show that the allowed states are $J = 0^+, 2^+, 3^+, 4^+$, and 6^+.

Q.9. How are the magnetic moments of nuclei in the shell model determined?
   a) What are the magnetic moments of a proton and a neutron?
   b) How does the calculated magnetic moment s of nuclei compare with the experimental values? Draw a diagram of magnetic moments of odd nuclei and its comparison with Schmidt lines.

Q.10.
   a) Describe the liquid drop model of a nucleus. Specify all the energy term which are relevant.
   b) How is this model used in calculating the binding energy of nuclei?

Q.11. The odd-even permanently deformed nucleus Y^{169} has ground state spin $7/2^+$ and first excited state spin $J^\pi = 1/2-$ and energies of excited states in keV as shown.

J π E (keV)	J^π	E (keV)	
		7/2— _____	?
? _____	?	? _____	204.0
		3/2— _____	99.0
9/2^+ _____	70.9		
		1/2— _____	24.3
7/2 + _____	0		

Calculate the energies and J^π values of states with ? marks.

8.5. CHAPTER 5

Q. 1. Why do some nuclei decay by α particle emission?
   a) Who discovered radioactivity, and name the radiation emitted and their properties.

b) How does radioactive nuclei decay, and what is half-life?

c) On what factors does the probability of such decay depend and why?

d) What is the relation between decay constant λ and ϑ-particle energy? Explain.

e) What are the approximate ranges of half-life of β-decays of nuclei?

f) What are the units in which radioactivity is expressed?

Q.2. U^{238} decays by α particle emission with a half-life of 10^3 years. If atomic masses of U^{238} and alpha particles are as 238.0289 and 4.0026 and mass of $Th^{23}4$ is 234.02,

a) calculate α particle energy,

b) calculate the ℓ-value of α particle,

c) determine the barrier penetration factor,

d) calculate the reduced width of decay.

Q.3.

a) Probability of α decay depends upon their energy and angular momentum (ℓ). Explain why.

b) How does the barrier penetration factor depend upon ℓ-values?

Q.4.

a) Draw a diagram showing the α particle potential barrier as a function of distance from the center.

b) Write the expression for wave number k as a function of alpha particle energy and nuclear and centrifugal potentials.

c) Write the expression for barrier penetration B as a function of k and distance (r) from the center.

Q. 5.

a) What is β decay, and when does a nucleus decay by β decay? Give examples of β decay.

b) β particle energies are observed to be continuous. Why was it considered a violation of energy conservation?

c) Who and what explanation was offered to explain this violation of energy?

d) What is a Kurie plot? Draw a diagram to show this for the decay of H^3 to He^3 with an end point energy of 18.1 keV.

Q. 6. Complete the following equations of β decays by writing down the missing information shown as X and Y.

a) $_6C^{14} \rightarrow e^- + X + Y$
b) $P \rightarrow n + X + Y$
c) $_8O^{14} \rightarrow X + e^+ + Y$
d) $_{38}Sr^{90} \rightarrow _{39}Y^{90} + X + Y$
e) Using atomic mass tables, calculate the energy released in each of these decays.

Q. 7. Nuclei can be excited to higher states by different processes.
a) Describe what are the processes by which a nucleus can be excited.
b) If a neutron of energy of 1 MeV is captured by a nucleus, what would be the energy of excited state of the compound nucleus?
c) How does this excited state decay?

Q.8. We know that probability of β decay depends upon many factors.
a) Describe these factors for β decay.
b) What do you mean by log ft values for β decay? What are f and t in this expression?
c) Give values of log ft for transitions, which are allowed and which are forbidden.
d) How does Coulomb potential affect the β spectrum for e⁻ and e+?

Q.9. β decay occurs from an initial state of a nucleus to a final state of the residual nucleus.
a) Show by a diagram how the initial state decays to final state of the nucleus.
b) What is the meaning of M_{if} and how is its value affected?
c) If one knows the value of M_{if} half-life and form factor for a given β decay, how can you determine the value of coupling constant G of β decay? Write the relation used.

Q.10. Classify the following decays according to the degree of forbiddeness.

a) Sr^{89} (5/2+) $\rightarrow$ Y^{89} (1/2-)
b) Cl^{36} (2+) $\rightarrow$ Ar^{36} (0+)

c) Al^{26} (5+)   → Mg^{26} (2+)
d) Si (0+)   → Mg^{26} (0+)
e) Zr^{97} (1/2+)   → Nb^{97} (1/2-)

Q.11. The following allowed β transitions are given.
    a)   Determine whether they are pure Fermi, pure Gamow-Teller, or a mixture of both types of interaction.

Transition	Spin-parity Initial Final		Log ft
$N^1 \rightarrow H^1$	½ +	½ +	3.07
$He^6 \rightarrow Li^6$	0 +	1 +	2.92
$O^{15} \rightarrow N^1$	½ -	½ -	3.64

Q.12.
    a)   When does a nucleus decay by gamma emission?
    b)   When current flows in an antenna wire, it generates an electromagnetic wave. Describe the nature and characteristic of this wave.
    c)   Gamma rays emitted from a nucleus are characterized as electronic or magnetic transitions. State the conditions for each type of decay.
    d)   The figure below shows a nucleus decay by emission of gamma rays labeled as 1, 2, 3, and 4. Calculate the energies and multipolarities of each transition.

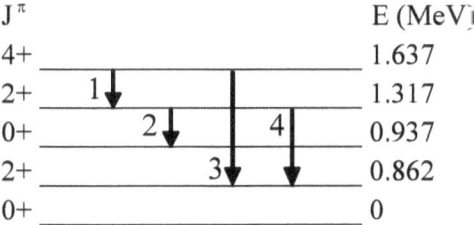

J^π		E (MeV)
4+	1↓	1.637
2+		1.317
0+	2↓ 4↓	0.937
2+	3↓	0.862
0+		0

Q.13. Discuss how the decay constant λ of γ rays for different multipolarities differ and how can you explain these in terms of shell and collective models?|

Q.14. A fictious nucleus has $J^{\pi} = \frac{1}{2}$ ground state and excited states of energies 0.50, 1.35, and 2.70 MeV, and $J^{\pi} = \frac{1}{2}+$, 3/2- and 5/2- respectively.
   a) Draw an energy level diagram.
   b) Calculate the γ-ray energies.
   c) Determine the multipolarities of each transition.
   d) Describe as to how the excited levels of this nucleus can be excited.

8.6. CHAPTER 6

Q.1. Describe a nuclear reaction. How is the Q-value of such a reaction determined? What quantities are conserved in a nuclear reaction?

Q.2. In the 1930s, neutron cross-section measurement showed sharp resonances, meaning large values as specific neutron energy. Discuss explanation given by N. Bohr for such resonances. Describe in detail the process and assumption made by him.

Q.3.
   a) Derive using partial wave analysis how the cross section σ for any value of ℓ can be expressed in terms of an amplitude and phase of the scattered wave.
   b) Discuss how the shape of resonance cross section looks for $\ell = 0$ and $\ell = 1$ resonances. (Draw a diagram.)

Q.4. According to compound nucleus, model resonances in cross section are due to some specific state to which the nucleus is excited.
   a) What does one know about the spacing distribution of these levels? Explain by drawing a diagram of level spacing distribution.

Q. 5 The measurements of neutron total cross section by Barshal et al. with poor resolution showed broad resonances at certain mass numbers for $\ell = 0$ and $\ell = 1$ resonance. Discuss who proposed a model to explain these broad resonances and describe some details of such a model.

Q.6. High resolution total neutron cross-section measurements allow us to determine a quantity known as strength function.
   a)   Explain what is the strength function, and how do you determine this quantity?
   b)   How does this quantity for $\ell = 0$ and $\ell = 1$ vary with atomic mass? (Draw a diagram.)

Q.7. Reactions of deuterons, He^3 and H^3 particles, with nuclei are classified as direct reactions. Describe these reactions, and discuss how the differential cross section varies as a function of angle for momentum transfer of $\ell = 0$ and $\ell = 2$. (Draw a diagram.)

Q.8. Several theories of shipping reactions have been proposed. Discuss the main concepts of these theories.

Q.9. Discuss photonuclear reactions and in particular photodisintegration of deuteron as a function γ-ray energies.

Q.10.
   a)   Discuss nuclear fission. What happens when a neutron strikes a nucleus of U^{235}?
   b)   What nuclei and nucleons are produced, and what is the energy released in the fission process?
   c)   How does the fission cross section in U^{235} vary as a function of neutron energy? (Draw a diagram to illustrate your discussion.)
   d)   Discuss the fission of U^{238} with neutrons.

Q.11.
   a)   Discuss fusion reaction between protons showing separate steps to reach the final $_2He^4$ nucleus.
   b)   How much energy is released in each step?

c) Why is it not practical to produce fusion energy on earth when it is taking place in the sun?

Q.12. Consider the induced nuclear reaction $_1H^2 + {_7}N^{14} \rightarrow {_6}C^{12} + {_2}He^4$. The atomic masses are $_1H^2$ (2.014 u), $_7N^{14}$(14.003 u), $_6C^{12}$(12.000u), and He^4(4.002u).
Calculate the energy released in this reaction.

Q.13. If 6.0 MeV deuterons strike a target of $_{19}K^{39}$,
   a) state the compound nucleus formed by the capture of the deuteron by the target and the excitation energy.
   b) how will the compound nucleus decay? Give different modes of decay by writing the equations for each type of decay.
   c) discuss on what factors such decays will depend.

Q.14. If after a deuteron strikes a target of Ti^{49} (ground state spin 7/2⁻) it emits a proton and leaves a residual nucleus Ti^{50} at the excited states with $J^\pi = 0^+, 2^+,$ and 3⁻.
   a) What angular momentum values are transferred to the residual nucleus by the capture of neutrons?
   b) For each value of (l) transferred, show by a diagram the angular distribution of protons.

Q.15. If a deuteron strikes a target Ti^{49} ($J^\pi = 7/2-$), it emits a proton and leaves the residual nucleus Ti^{50} in excited states with $J\pi = 0 +, 2+,$ and 3-.
   a) Describe what angular momentum values are transferred to the nucleus by the captured neutron.
   b) For each value of l of the captured neutron, show, by a diagram the angular distribution as a function of angle Θ of protons?

Q.16. Discuss the photonuclear reaction of deuteron.
   a) How does this cross section varies with the photon energy?
   b) What types of gamma rays are mainly responsible for the disintegration of the deuteron?

Q.17. In his compound nucleus theory, Bohr made an important assumption.
a) Discuss this assumption.
b) How and who performed some experiments to prove the validity of this assumption? Describe that experiment.

Q 18. Discuss the main features of the optical model.
a) What experimental data prompted the origin of this model, and how was it applied to explain the data?
b) What is meant by the neutron strength function, and how does this model explain the experimental data of neutron strength functions of $l = 0$ and $l = 1$ resonances.

8.7. CHAPTER 7

Q.1. Describe the various techniques that are currently used for taking images of the organs of human body.

Q.2. Describe the NMR technique and the nuclear principle involved in this technique to take clear images of the human body.

Q.3. Describe the techniques that are currently used in the treatment of tumors and other cancerous growths in the human body.
Discuss the use of radioactive isotopes and particle accelerators for such treatment.

Q.4. Nuclear fission of uranium is used to produce nuclear energy.
a) Describe the different types of nuclear reactors, which are in use in the world.
b) Discuss the merits of each of these reactors.

Q.5. A moderator is used in a nuclear reactor.
Discuss the need for a moderator as why it is used.
a) Describe different types of moderators pointing out the advantages and disadvantages of each of these moderators.
b) How is the power generated in a reactor controlled?

Q.6. U^{238} and Th^{232} are found in great abundance in our planet.
  a)   How can one use these materials to produce nuclear energy?
  b)   Discuss the nuclear reactions to produce fissionable nuclei.
  c)   What type of nuclear reactor is used to generate nuclear energy from these nuclei and why?

Q.7. Describe the neutron activation analysis known as NAA to determine the traces of elements found in our environment and in the detection of crime.

Q.8. Describe the nuclear reactions occurring in the interior of our sun to produce energy.
  a)   Discuss how the heavier mass nuclei are formed in high-mass stars.
  b)   Describe how the abundance of heavier nuclei were created in our planet.
  c)   Starting from hydrogen, describe the formation of heavier nuclei, the temperatures needed, and duration of such reaction in high-mass stars.

8.8. TESTING OF STUDENT'S KNOWLEDGE IN NUCLEAR PHYSICS

This book is written for a one-semester course on the basic concepts of fundamental topics in nuclear physics. Such a course is offered to seniors and beginning graduate students pursuing a BS, MS, or a PhD degree in physics or nuclear engineering.

The student's knowledge of this subject is tested in a series of tests given at regular interval during the class. These tests are based upon similar or same problems given in chapter 8 of this book. The answers of these problems are not given, but these would be discussed in the class.

The first test will be on the basic properties of nuclei as discussed in chapter 1.

The second test will be on the nature of nuclear force as obtained from deuteron and from pairs of nucleon-nucleon interaction as are discussed in chapters 2 and 3.

The third test will be on nuclear structure learned from different nuclear models as are discussed in chapter 4.

The fourth test will be on modes of nuclear decays as are discussed in chapter 5.

The fifth test will be on types of nuclear reactions as are discussed in chapter 6.

At the end of the semester, a comprehensive final exam based upon all the topics discussed in the textbook will be given.

AUTHOR'S REFERENCES

Chapter 1

(1.1)   Rutherford, E., Phil Mag. 21, 669 (1911) Proc. Roy Soc (London) A 81, 141 (1908)

(1.2)   Chadwick, J., Proc. Roy. Soc (London) A 136, 692 (1932)

(1.3)   Hofstadter, R., Fletcher H. R., McIntyre J. A., Phys Rev 92, 978 (1953)

(1.4)   Hahn, B.D., Ravenhall, D. G. and Hofstadter, R. Phys Rev 101, 131 (1956)

(1.5)   Lauritsen,T.S. Ajzenberg.Seleve, F., Nuclear Data Sheets (1962)

Chapter 2

(2.1)   Yukawa, Heidi, Proc. Phy. Mat. Soc (Japan), 17, 48, (1935)

(2.2)   Powell C. F. and O'Chiliani G., Phys.Soc.(London)150,(1947)

(2.3)   Gell-Mann, M., Phys Rev 125, 1067, (1962)

(2.4)   McIntyre, J. A., and Hofstadter, R. Phys Rev 98 158 (1955), McIntyre JA Burleson G.R. Phys Rev 112, 2077 (1958)

(2.5)   Pauli, W Die Allegemeinen Princizipien des Wellen mechanick Haudbuch der Physik, vol 25, 1 (1933)

Chapter 3

(3.1)   Storrs, S. C. I. and Frisch, D. H., Phys Rev 95, 1252 (1954)

(3.2)   Van de Graff, R.J. Phys Rev 38, 1931 (1919)

(3.3)   Adair, R. K. Rev. Mod. Phys 22, 249 (1950)

(3.4)   Knecht, D. J., Messel S., Benrers E. D. and Northcliff. C. Phys. Rev 114, 550, (1950)

(3.5)   Breit, G. et al. Phys Rev. 120, 2227 (1960)

(3.6)   Hull, M. H. et. al. Phys Rev 122 1606, (1961)

(3.7)   Yang, C. N. and Wolfenstein, L. Phys Rev 74, 764 (1948) and Phys Rev 75, 1664 (1949)

(3.8)   Gell-Mann M. Phys. Rev. 125,1067 (1962)

(3.9)   Gell-Mann M. Phy. Lett. 8,214 (1964)

Chapter 4

(4.1) Aston, F. Phil Mag (London) 38, 707 (1919)

(4.2) Fermi, E. Z. Physik 88, 161 (1934), Phy Rev 48, 570 (1935)

(4.3) Mayer, M. G. Phys. Rev 75, 1969 (1949) and Haxel, O., Jensen J. H., Suess H. E. Phys Rev 75, 1766 (1949)

(4.4) Dicke R. H. and Wittke J. P. Introd. To quantum mechanics Addison Wesley (1960)

(4.5) Feld, B. T. Ann Rev Nuc Sci 2, 249 (1953)

(4.6) Schmidt, T. Z. Physik,106, 358 (1937)

(4.7) Segre, E. Nuclei and Particles NY WA Benjamin In (1964)

(4.8) Rainwater, James, Phys Rev 79, 432 (1950)

(4.9) Bohr A. Mottelson B. R. Phy. Rev. 89, 316 (1953a)

(4.10) Irvine J. M. Nuclear structure theory Pergaman Press (1972)

(4.11) Bohr, A. States of Atomic Nuclei, Mat. Fys.Skr. Dan Vid, selsk, 1 No. 8 (1959)

(4.12) Nathan, O. and Nilsson, S. G.,(Ed.) by Siegbahn, K., beta & γ-ray spectroscopy, North Holland Pub. Co., Amsterdam (1959)

(4.13) Harmatz, B., Handley T. H., and Mihelich J. W. Phys Rev 114 1082 (1959)

(4.14) Wilson, R. G. and Pool, M. L. Phy. Rev. 120,1843 (1960)

(4.15) Eichler, E., Rev of Mod Physics 36 809 (1964)

(4.16) Nilsson, S. G. Mat. Fys. Medd. Dan. Vid. Selsk 29, 228 (1955)

(4.17) Mottelson, B. R. and Nilsson, S. G. Mat Fys. Skv. Dan. Vid. Selsk 1. No 8 (1959)

(4.18) Bohr, N. & Wheeler, J. A. Phys Rev 56, 429 (1939)

(4.19) Weizsaker, Von, C. F. Z Physik 96, 421 (1935)

Chapter 5

(5.1) Becquerel H. Compt Rend Paris 122, 420 (1896)

(5.2) Geiger H. and Nuttall J. M. Phil. Mag. 22, 613 (1911)

(5.3) Stephens F. S. Nuclear Spectroscopy Part A Ed. By Fay Ajzenberg-Selov, NY Academic Press

(5.4) Igo G. Phys Rev lefters 1, 72 (1958)

(5.5) Chadwick J. Proc Roy Soc (Landon) A 136, 692 (1932)

(5.6) Neary G. J. Proc. Phy. Soc. (London) A 175, 71 (194

(5.7) Pauli W. Dae Allyemeien Pennazipien der Wellenmechanick Handbuch der Physik vol 24, 1, (1933)

(5.8) Reines F. and Cowan C. L., Phy Rev 113, 273 (1959)

(5.9) Dirac P.A.M. Proc Roy Soc 126, 360 (1930) AND 133, 61 (1931)

(5.10) Fermi E. Z. Physik 88, !61 (1934)

(5.11) Reitz J.R. Phys Rev 77, 50 (1950)

(5.12) Kurie E N. D. Richardson J.R. and Paxton H.C. Phys Rev 49, 368 (1936)

(5.13) Langer, L. M. and Moffat R. D. Phy. Rev 88, 689 (1952)

(5.14) Feenberg, E. and Trigg, G. Rev of Mod Physics 22, 399 (1950)

(5.15) Lee T. D. and Yang C. N. Phys Rev 119, 1410 (1960)

(5.16) Wu C. S. Ambler E. Hayward, R.W., Hoppes D. D. and Hudson R. Phys. Rev. 105 1413 (1957)

(5.17) Condon E. U. And Odishaw H. Handbook of Physics McGraw Hill, New York (1958)

(5.18) Weisskopf,V. F., Phy.Rev. 83,1073,(1951)

(5.19) Yukawa H. S. Sakata S. Proc. Phys Soc. Japan, 17, 467 (1935)

(5.20) Alburger D. E. and Hedgran A. Arkin Fysik 7, 424 (1953-54)

(5.21) Evan,G.T. and Graham,G.T. in α,β andγ-ray spectroscopy ed. by Siegbahn,962 (1965)

Chapter 6

(6.1) Cockroft, J. D. and Walton E. T. S. Proc Roy Soc 137, 229 (1929)

(6.2) Van de Graaff, R. J. Phys Rev 28, 1919 (1931)

(6.3) Lawrence, E. O. and Livingston, M. S. Phys Rev 37, 1707 (1931)

(6.4) Van de Graaff Nucl. Inst. Methods 8,!95 (1960)

(6.5) Breit, G. and Wigner P., Phys Rev. 48 918, (1935)

(6.6) Garg J. B. Tikku, V. K., Harvey, J. Halperin J., and Macklin R. L. Phy. Rev. C,1808 (1981)

(6.7) Bohr, N. Nature 137, 344 (1936)

(6.8) Ghoshal, S. N., Phys Rev 80, 939 (1950)

(6.9) Breit, G. and Wigner, E. Phy Rev 49, 519 (1936)

(6.10) Kapur P. L. & Peierls, R. E. Proc. Roy. Soc. (London) A 116 277 (1938)

(6.11) Wigner E. and Eisenbud L. Phy. Rev. 72, 29 (1947)

(6.12) Garg, J. B., Rainwater J., Petersen J. S., and Haven W. W. Jr., Phys Rev 134 B985 (1964)

(6.13) Wigner E. P. Proc of Conf. On neutron physics, Gatlinburg, ORNL-2309(1957)

(6.14) Mehta M. L. and Dyson, F. J., J. Math. Phys, 713 (1963)

(6.15) Garg J. B. Rainwater J. and Havens W.W. Jr. Phy.Rev. 137, B547 (1965)

(6.16) Bilpuch et al. Proc. Statistical Prop. Nuclei Plenum Press (1972)

(6.17) Porter, C. E. and Thomas R. G. Phys Rev 104 483 (1956)

(6.18) Kaufman G., Goldberg E., Koster L. J., & Mooring F. P. Phy Rev. 88, 675 (1952)

(6.19) Barshall, H. H. Phys Rev 76 1146 (1949)

(6.20) Feshbach, H., Porter, C. E., and Weisskopf, V. F. Phys Rev. 96, 448 (1954)

(6.21) Farrel J. A., Kyker G. C., Bilpuch E. G., and Newson H. Phy letters 17, 286 (1965)

(6.22) Oppenheimer J. R. and Philips M. Phy Rev 48, 500 (1935)

(6.23) Butler, S. T. Proc Roy Soc. A 208, 559 (1951)

(6.24) Rapaport, J. Sperduto, A. and Buechner, W. W. Phy. Rev. 151, 939 (1966)

(6.25) Tobocman, W. Phy. Rev. 94,1655 (1954)

(6.26) Huby R., Refai M. Y., and Satchler G. R. Nucl Phy 9, 94 (1958)
         (6.27) Bhatia, A. B., Huang, K. Huby, R., Newns, H. C. Phil mag 43, 485 (1952)

(6.28) Bethe H. A. and Butler S. T., Phy Rev 85 1045 (1952)

(6.29) Kikuchi, K. Kabayashi, S. and Matsuda, K. J. Phy. Soc. Japan 14,121 (1959)

(6.30) Evans R. D. The Atomic Nucleus NY McGraw-Hill Book Co. (1955)

(6.31) Bramblett, R. L., Caldwell, J. T., Harvey R., and Fultz, S.C. Phys Rev 133, B869 (1964)

(6.32) Hahn, O. and Strassman, F. Die Naturewersenschaften, 277, 11, (1939)

(6.33) Meitner, L. and Frisch, O. R., Nature 143 239 (1939)

(6.34) Bohr, N. and Wheeler, J. A., Phys. Rev. 56, 426, (1939)

(6.35) Fong P. Phy. Rev. 89, 332 (1953)

(6.36) Hughes, D. J. and Schwartz, R. B. Neutron Cross Section BNL Report 325 second edition (1958)

(6.37) Block B. and Feshbach H. Annals of Physics 23,47 (1963)

(6.38) Wilenzick R. M., Mitchel G., Seth K. K., and Lewis,H. W. Phy. Rev. 121,1150 (1961)

(6.39) Gale,N.H., Garg,J.B., Calvert, J.& Ramavataram,K. Nucl. Phy. 20,313(1960)

(6.40) Newns H. C. (1960) to be published

(6.41) Forsyth F. D., Barros, F. de S., Jaffe, A. A., Taylor, J. & Ramavataram, S., Proc. Phy. Soc. 75,291 (1960)

(6.42) Toohey R. E., and Jackson H. E. Phy. Rev.C 1440 (1972)

(6.43) Eisenberg, R. M. & Porter C. E., Rev. of Mod. Phy. 33,219 (1961)

(6.44) Fowler J. L. & Rosen L. Phy.Rev. 72, 926 (1947)

(6.45) Cao M. G., Migneco E., Theobald P. Wartena J.A. and Winter J., Journ. Of Nuclear Energy 22,211(1967)

(6.46) Wapstra A. H. & Audi G. Nucl.Phy. A 432,1, (1985)

(6.47) Zeidman, Z. and Yntema,J. L.Nucl. Phy.12,298 (1959)

OTHER PUBLISHED BOOKS
ON THE TOPIC OF NUCLEAR PHYSICS.

1. Bertulani, C. A. *Nuclear Physics in a Nutshell*. Princeton University Press, 2007.
2. B. R. Martin, Nuclear and Particle physics, Publ. John Wiley and sons (2006)
3. K. Hyde, Basic ideas and concepts in nuclear physics (An introduction) pub. Institute of physics (2004)
4. A. Das and T. Ferbel, Introduction to nuclear and particle physics. Pub. World Scientific (2003-05)
5. W. N. Cottingham and D.A. Greenwood, Intod. to nuclear physics Cambridge University press (1986-2001)
6. R.A. Dunlap, The physics of Nuclei and particles, Thomson Learning Book/Cole (2004)
7. J.Lilley Nuclear physics principles and applications, Pub. John Willey & sons (2001)
8. P.E. Hodgson, E. Gadioli and E. Gadioli Erba, Introd. Nuclear Physics, Oxford University Press (1997)
9. W.S.C. Williams Nuclear & particle physics, Pub. Oxford University press (1991)
10. N. A. Jelly, Fundamentals of nuclear physics, Cambridge University press (1990)
11. S.M. Wong, Introductory nuclear physics Pub. Prentice Hall (1990) 2nd Edition John Wiley and sons (1998)
12. K.S. Krane, Introduction to nuclear physics, Pub. John Wiley & sons (1987)
13. J.M. Pearson, Nuclear physics, energy and matter, Pub. Adam Hilger (1986)
14. E. Segre, Nuclei and particles, W.A. Benjamin (1977)
15. A. Bohr and B. R. Motteslson, Nuclear structure Vol 1 &2, Pub. W.A. Benjamin Inc. (1975)

16. W.E. Burcham, Nuclear physics an introduction, Pub. Longman (London) (1973)
17. B. L. Cohen, Concepts of nuclear physics, Pub. McGraw Hill (!971)
18. Von Butler, Nuclear physics an introduction, Pub. Academic press! 1968)
19. R.R.Roy and B.P. Nigam, Nuclear physics, Pub. John Wiley & sons (1967)
20. P. Marimer and E.Sheldon, Physics of nuclei and particles, pub. Academic press Vol 1&2 (1969)\
21. H.A. Enge, Introduction to nuclear physics, Pub. Addison Wesley (1966)
22. A.M. Preston, Physics of the nucleus Pub. Addison Wesley (1962)
23. I. Kaplan, Nuclear physics, Pub. Addison Wesley (1955)
24. R.D. Evans, The Atomic Nucleus Pub. McGraw Hill N.Y. (1955)
25. A.E.S. Green, Nuclear physics Pub. McGraw Hill (1955)
26. J.M. Blatt & V. Weisskopf, Theoretical Nuclear physics, Pub. John Willey & sons(1952) 2nd ed. Pub. Springer & Verlog (1979)

Books on special topics

G. E. Brown &A. D. Jackson Nucleon-nucleon interaction Pub. North Holland Co. (1976)
Green Swada and Saxton. Nuclear independent particle model, Pub. Academic press (1968)
P.M. Endt and M. Demeur -Nuclear Reactions vol I Pub. North Holland Pub. Co. (1959)
S.T. Butler- Nuclear stripping reactions pub. John Wiley & sons(1957)
W.F. Hornyack- Nuclear structure Pub. Academic press (1975)
D.J. Rowe- Nuclear Collective motion Pub. Methuen (1970)
K. Siegbahn- Beta and gamma ray spectroscopy
Pub.North Holland Co. (1955)
Meyer and J.H.D Jensen, Elementary theory of nuclear shell structure, Pub. John Wiley and Sons (1957)
I.E. McCarthy, Introd.to Nuclear Theory, Pub. John Wiley & sons (1968)
Donald H. Perkins, Introd. To High energy Physics, Pub.Cambridge Unibversity Press. (2000)

TABLE OF ATOMIC MASSES, ABUNDANCES AND HALF-LIFE OF DECAY

The following table gives the atomic masses in (amu), J^π values abundances (percentage) and unstable half-life of decay and principal modes of decay such as α, β, and γ-ray decays.

Masses are those of neutral atoms from 1983 atomic mass evaluation by A. H. Wapstra and G. Andi, Nucl. Phys. A 432, 1 (1985).The table used in this book is taken from the book Introd. to nuclear physics by K.S. Krane published by John Wiley & sons (1987). Permission was granted.

This table is useful in calculating the Q-values of a nuclear reaction or nuclear decay.

	Z	A	Atomic mass (u)	J^π	Abundance or Half-life
H	1	1	1.007825	$\frac{1}{2}^+$	99.985%
		2	2.014102	1^+	0.015%
		3	3.016049	$\frac{1}{2}^+$	12.3 y (β^-)
He	2	3	3.016029	$\frac{1}{2}^+$	1.38×10^{-4}%
		4	4.002603	0^+	99.99986%
Li	3	6	6.015121	1^+	7.5%
		7	7.016003	$\frac{3}{2}^-$	92.5%
		8	8.022486	2^+	0.84 s (β^-)
Be	4	7	7.016928	$\frac{3}{2}^-$	53.3 d (ε)
		8	8.005305	0^+	0.07 fs (α)
		9	9.012182	$\frac{3}{2}^-$	100%
		10	10.013534	0^+	1.6 My (β^-)
		11	11.021658	$\frac{1}{2}^+$	13.8 s (β^-)
B	5	8	8.024606	2^+	0.77 s (ε)
		9	9.013329	$\frac{3}{2}^-$	0.85 as (α)
		10	10.012937	3^+	19.8%
		11	11.009305	$\frac{3}{2}^-$	80.2%
		12	12.014353	1^+	20.4 ms (β^-)
		13	13.017780	$\frac{3}{2}^-$	17.4 ms (β^-)
C	6	9	9.031039	$\frac{3}{2}^-$	0.13 s (ε)
		10	10.016856	0^+	19.2 s (ε)
		11	11.011433	$\frac{3}{2}^-$	20.4 m (ε)
		12	12.000000	0^+	98.89%
		13	13.003355	$\frac{1}{2}^-$	1.11%
		14	14.003242	0^+	5730 y (β^-)
		15	15.010599	$\frac{1}{2}^+$	2.45 s (β^-)
N	7	12	12.018613	1^+	11 ms (ε)
		13	13.005739	$\frac{1}{2}^-$	9.96 m (ε)
		14	14.003074	1^+	99.63%
		15	15.000109	$\frac{1}{2}^-$	0.366%
		16	16.006100	2^-	7.13 s (β^-)

Z	A	Atomic mass (u)	I^π	Abundance or Half-life
	17	17.008450	½⁻	4.17 s (β^-)
	18	18.014081	1⁻	0.63 s (β)
O 8	14	14.008595	0⁺	71 s (ε)
	15	15.003065	½⁻	122 s (ε)
	16	15.994915	0⁺	99.76%
	17	16.999131	5/2⁺	0.038%
	18	17.999160	0⁺	0.204%
	19	19.003577	5/2⁺	26.9 s (β)
	20	20.004076	0⁺	13.5 s (β^-)
F 9	17	17.002095	5/2⁺	64.5 s (ε)
	18	18.000937	1⁺	110 m (ε)
	19	18.998403	½⁻	100%
	20	19.999981	2⁺	11 s (β^-)
	21	20.999948	5/2⁺	4.3 s (β^-)
	22	22.003030	(3,4)⁺	4.2 s (β^-)
	23	23.003600	(½,5/2)⁺	2.2 s (β^-)
Ne 10	17	17.017690	½	0.11 s (ε)
	18	18.005710	0⁺	1.7 s (ε)
	19	19.001880	½⁺	17.3 s (ε)
	20	19.992436	0⁺	90.51%
	21	20.993843	3/2⁺	0.27%
	22	21.991383	0⁺	9.22%
	23	22.994465	5/2⁺	37.6 s (β)
	24	23.993613	0⁺	3.4 m (β^-)
	25	24.997690	(½,3/2)⁺	0.60 s (β^-)
Na 11	20	20.007344	2⁺	0.45 s (ε)
	21	20.997651	3/2⁺	22.5 s (ε)
	22	21.994434	3⁺	2.60 y (ε)
	23	22.989768	3/2⁺	100%
	24	23.990961	4⁺	15.0 h (β^-)
	25	24.989953	5/2⁺	60 s (β)
	26	25.992586	3⁺	1.1 s (β^-)
	27	26.993940	5/2⁺	0.30 s (β)
Mg 12	21	21.011716	(3/2,5/2)⁺	0.123 s (ε)
	22	21.999574	0⁺	3.86 s (ε)
	23	22.994124	3/2⁺	11.3 s (ε)
	24	23.985042	0⁺	78.99%
	25	24.985837	5/2⁺	10.00%
	26	25.982594	0⁺	11.01%
	27	26.984341	½⁺	9.46 m (β^-)
	28	27.983877	0⁺	21.0 h (β^-)
	29	28.988480	3/2⁺	1.4 s (β)
Al 13	24	23.999941	4⁺	2.07 s (ε)
	25	24.990429	5/2⁻	7.18 s (ε)

Z	A	Atomic mass (u)	I^π	Abundance or Half-life
	26	25.986892	5⁺	0.72 My (ε)
	27	26.981539	5/2⁺	100%
	28	27.981910	3⁺	2.24 m (β^-)
	29	28.980446	5/2⁺	6.6 m (β)
	30	29.982940	3⁻	3.7 s (β^-)
Si 14	26	25.992330	0⁺	2.21 s (ε)
	27	26.986704	5/2⁺	4.13 s (ε)
	28	27.976927	0⁺	92.23%
	29	28.976495	½⁺	4.67%
	30	29.973770	0⁺	3.10%
	31	30.975362	3/2⁺	2.62 h (β^-)
	32	31.974148	0⁺	105 y (β^-)
	33	32.997920	(3/2⁺)	6.2 s (β^-)
P 15	29	28.981803	½⁺	4.1 s (ε)
	30	29.978307	1⁺	2.50 m (ε)
	31	30.973762	½⁺	100%
	32	31.973907	1⁺	14.3 d (β)
	33	32.971725	½⁺	25.3 d (β^-)
	34	33.973636	1⁺	12.4 s (β^-)
S 16	30	29.984903	0⁻	1.2 s (ε)
	31	30.979554	½⁺	2.6 s (ε)
	32	31.972071	0⁺	95.02%
	33	32.971458	3/2⁺	0.75%
	34	33.967867	0⁺	4.21%
	35	34.969032	3/2⁺	87.4 d (β^-)
	36	35.967081	0⁺	0.017%
	37	36.971126	5/2⁻	5.0 m (β^-)
	38	37.971162	0⁺	170 m (β^-)
Cl 17	33	32.977452	½⁺	2.51 s (ε)
	34	33.973763	0⁺	1.53 s (ε)
	35	34.968853	3/2⁺	75.77%
	36	35.968307	2⁺	0.30 My (β^-)
	37	36.965903	3/2⁺	24.23%
	38	37.968011	2⁻	37.3 m (β^-)
	39	38.968005	3/2⁺	56 m (β)
	40	39.970440	2⁻	1.35 m (β^-)
	41	40.970590	(½,3/2)⁻	31 s (β^-)
Ar 18	34	33.980269	0⁺	0.844 s (ε)
	35	34.975256	3/2⁺	1.78 s (ε)
	36	35.967546	0⁺	0.337%
	37	36.966776	3/2⁺	35.0 d (ε)
	38	37.962732	0⁺	0.063%
	39	38.964314	7/2⁻	269 y (β^-)
	40	39.962384	0⁺	99.60%
	41	40.964501	7/2⁻	1.83 h (β^-)

Z	A	Atomic mass (u)	I^π	Abundance or Half-life		Z	A	Atomic mass (u)	I^π	Abundance or Half-life
	42	41.963050	0^+	33 y (β^-)			52	51.946898	0^-	1.7 m (β^-)
	43	42.965670		5.4 m (β^-)			53	52.949730	$(\frac{3}{2})^-$	33 s (β^-)
	44	43.965365	0^+	11.9 m (β^-)		V 23	46	45.960198	0^+	0.42 s (ε)
K 19	37	36.973377	$\frac{3}{2}^+$	1.23 s (ε)			47	46.954906	$\frac{3}{2}^-$	32.6 m (ε)
	38	37.969080	3^+	7.61 m (ε)			48	47.952257	4^+	16.0 d (ε)
	39	38.963707	$\frac{3}{2}^+$	93.26%			49	48.948517	$\frac{7}{2}^-$	330 d (ε)
	40	39.963999	4^-	1.28 Gy (β^-)			50	49.947161	6^+	0.250%
	41	40.961825	$\frac{3}{2}^+$	6.73%			51	50.943962	$\frac{7}{2}^-$	99.750%
	42	41.962402	2^-	12.4 h (β^-)			52	51.944778	3^+	3.76 m (β^-)
	43	42.960717	$\frac{3}{2}^-$	22.3 h (β^-)			53	52.944340	$\frac{7}{2}^-$	1.6 m (β^-)
	44	43.961560	2^-	22.1 m (β^-)			54	53.946442	$(3,4,5)^+$	50 s (β^-)
	45	44.960696	$\frac{3}{2}^+$	17 m (β^-)		Cr 24	46	45.968360	0^+	0.26 s (ε)
	46	45.961976	(2^-)	115 s (β^-)			47	46.962905	$\frac{3}{2}^-$	0.51 s (ε)
	47	46.961677	$\frac{1}{2}^+$	17.5 s (β^-)			48	47.954033	0^+	21.6 h (ε)
Ca 20	38	37.976318	0^+	0.44 s (ε)			49	48.951338	$\frac{5}{2}^-$	41.9 m (ε)
	39	38.970718	$\frac{3}{2}^+$	0.86 s (ε)			50	49.946046	0^+	4.35%
	40	39.962591	0^+	96.94%			51	50.944768	$\frac{7}{2}^-$	27.7 d (ε)
	41	40.962278	$\frac{7}{2}^-$	0.10 My (ε)			52	51.940510	0^+	83.79%
	42	41.958618	0^+	0.647%			53	52.940651	$\frac{3}{2}^-$	9.50%
	43	42.958766	$\frac{7}{2}^-$	0.135%			54	53.938882	0^+	2.36%
	44	43.955481	0^+	2.09%			55	54.940842	$\frac{3}{2}^-$	3.50 m (β^-)
	45	44.956185	$\frac{7}{2}^-$	165 d (β^-)			56	55.940643		5.9 m (β^-)
	46	45.953689	0^+	0.0035%		Mn 25	50	49.954240	0^+	0.28 s (ε)
	47	46.954543	$\frac{7}{2}^-$	4.54 d (β^-)			51	50.948213	$\frac{5}{2}^-$	46.2 m (ε)
	48	47.952533	0^+	0.187%			52	51.945568	6^+	5.59 d (ε)
	49	48.955672	$\frac{3}{2}^-$	8.72 m (β^-)			53	52.941291	$\frac{7}{2}^-$	3.7 My (ε)
	50	49.957519	0^+	14 s (β^-)			54	53.940361	3^+	312 d (ε)
Sc 21	42	41.965514	0^+	0.68 s (ε)			55	54.938047	$\frac{5}{2}^-$	100%
	43	42.961150	$\frac{7}{2}^-$	3.89 h (ε)			56	55.938907	3^+	2.58 h (β^-)
	44	43.959404	2^+	3.93 h (ε)			57	56.938285	$\frac{5}{2}^-$	1.6 m (β^-)
	45	44.955910	$\frac{7}{2}^-$	100%			58	57.940060	3^+	65 s (β^-)
	46	45.955170	4^+	83.8 d (β^-)		Fe 26	51	50.956825	$(\frac{5}{2})^-$	0.25 s (ε)
	47	46.952409	$\frac{7}{2}^-$	3.35 d (β^-)			52	51.948114	0^+	8.27 h (ε)
	48	47.952235	6^+	43.7 h (β^-)			53	52.945310	$\frac{7}{2}^-$	8.51 m (ε)
	49	48.950022	$\frac{7}{2}^-$	57.0 m (β^-)			54	53.939613	0^+	5.8%
	50	49.952186	5^+	1.71 m (β^-)			55	54.938296	$\frac{3}{2}^-$	2.7 y (ε)
Ti 22	43	42.968523	$\frac{7}{2}^-$	0.51 s (ε)			56	55.934939	0^+	91.8%
	44	43.959690	0^+	54 y (ε)			57	56.935396	$\frac{1}{2}^-$	2.15%
	45	44.958124	$\frac{7}{2}^-$	3.09 h (ε)			58	57.933277	0^+	0.29%
	46	45.952629	0^+	8.2%			59	58.934877	$\frac{3}{2}^-$	44.6 d (β^-)
	47	46.951764	$\frac{5}{2}^-$	7.4%			60	59.934078	0^+	1.5 My (β^-)
	48	47.947947	0^+	73.7%			61	60.936748	$(\frac{3}{2},\frac{5}{2})$	6.0 m (β^-)
	49	48.947871	$\frac{7}{2}^-$	5.4%			62	61.936773	0^+	68 s (β^-)
	50	49.944792	0^+	5.2%		Co 27	54	53.948460	0^+	0.19 s (ε)
	51	50.946616	$\frac{3}{2}^-$	5.80 m (β^-)						

Z	A	Atomic mass (u)	I^π	Abundance or Half-life		Z	A	Atomic mass (u)	I^π	Abundance or Half-life
	55	54.942001	$\frac{7}{2}^-$	17.5 h (ε)		Ga 31	64	63.936836	0^+	2.6 m (ε)
	56	55.939841	4^+	78.8 d (ε)			65	64.932738	$\frac{3}{2}^-$	15.2 m (ε)
	57	56.936294	$\frac{7}{2}^-$	271 d (ε)			66	65.931590	0^+	9.4 h (ε)
	58	57.935755	2^+	70.8 d (ε)			67	66.928204	$\frac{3}{2}^-$	78.3 h (ε)
	59	58.933198	$\frac{7}{2}^-$	100 %			68	67.927982	1^+	68.1 m (ε)
	60	59.933820	5^+	5.27 y (β^-)			69	68.925580	$\frac{3}{2}^-$	60.1 %
	61	60.932478	$\frac{7}{2}^-$	1.65 h (β^-)			70	69.926028	1^+	21.1 m (β^-)
	62	61.934060	2^+	1.5 m (β^-)			71	70.924701	$\frac{3}{2}^-$	39.9 %
	63	62.933614	$(\frac{7}{2})^-$	27.5 s (β^-)			72	71.926365	3^-	14.1 h (β^-)
							73	72.925169	$\frac{3}{2}^-$	4.87 h (β^-)
Ni 28	55	54.951336	$\frac{7}{2}^-$	0.19 s (ε)			74	73.926940	$(4)^-$	8.1 m (β^-)
	56	55.942134	0^+	6.10 d (ε)			75	74.926499	$\frac{3}{2}^-$	2.1 m (β^-)
	57	56.939799	$\frac{3}{2}^-$	36.0 h (ε)						
	58	57.935346	0^+	68.3 %		Ge 32	66	65.933847	0^+	2.3 h (ε)
	59	58.934349	$\frac{3}{2}^-$	0.075 My (ε)			67	66.932737	$(\frac{1}{2})^-$	19.0 m (ε)
	60	59.930788	0^+	26.1 %			68	67.928096	0^+	271 d (ε)
	61	60.931058	$\frac{3}{2}^-$	1.13 %			69	68.927969	$\frac{5}{2}^-$	39.0 h (ε)
	62	61.928346	0^+	3.59 %			70	69.924250	0^+	20.5 %
	63	62.929670	$\frac{1}{2}^-$	100 y (β^-)			71	70.924954	$\frac{1}{2}^-$	11.2 d (ε)
	64	63.927968	0^+	0.91 %			72	71.922079	0^+	27.4 %
	65	64.930086	$\frac{5}{2}^-$	2.52 h (β^-)			73	72.923463	$\frac{9}{2}^+$	7.8 %
	66	65.929116	0^+	54.8 h (β^-)			74	73.921177	0^+	36.5 %
	67	66.931570	?	21 s (β^-)			75	74.922858	$\frac{1}{2}^-$	82.8 m (β^-)
							76	75.921402	0^+	7.8 %
Cu 29	59	58.939503	$\frac{3}{2}^-$	82 s (ε)			77	76.923548	$\frac{7}{2}^+$	11.3 h (β^-)
	60	59.937366	2^+	23.4 m (ε)			78	77.922853	0^+	1.45 h (β^-)
	61	60.933461	$\frac{3}{2}^-$	3.41 h (ε)			79	78.925360	$(\frac{1}{2})^-$	19 s (β^-)
	62	61.932586	1^+	9.73 m (ε)						
	63	62.929599	$\frac{3}{2}^-$	69.2 %		As 33	70	69.930929	4^+	53 m (ε)
	64	63.292766	1^+	12.7 h (ε)			71	70.927114	$\frac{5}{2}^-$	61 h (ε)
	65	64.927793	$\frac{3}{2}^-$	30.8 %			72	71.926755	2^-	26.0 h (ε)
	66	65.928872	1^+	5.10 m (β^-)			73	72.923827	$\frac{3}{2}^-$	80.3 d (ε)
	67	66.927747	$\frac{3}{2}^-$	61.9 h (β^-)			74	73.923928	2^-	17.8 d (ε)
	68	67.929620	1^+	31 s (β^-)			75	74.921594	$\frac{3}{2}^-$	100 %
							76	75.922393	2^-	26.3 h (β^-)
Zn 30	61	60.939514	$\frac{3}{2}^-$	89 s (ε)			77	76.920646	$\frac{3}{2}^-$	38.8 h (β^-)
	62	61.934332	0^+	9.2 h (ε)			78	77.921830	(2^-)	91 m (β^-)
	63	62.933214	$\frac{3}{2}^-$	38.1 m (ε)			79	78.920946	$\frac{3}{2}^-$	9.0 m (β^-)
	64	63.929145	0^+	48.6 %						
	65	64.929243	$\frac{5}{2}^-$	244 d (ε)		Se 34	71	70.932270	$\frac{5}{2}^-$	4.7 m (ε)
	66	65.926035	0^+	27.9 %			72	71.927110	0^+	8.4 d (ε)
	67	66.927129	$\frac{5}{2}^-$	4.10 %			73	72.926768	$\frac{9}{2}^+$	7.1 h (ε)
	68	67.924846	0^+	18.8 %			74	73.922475	0^+	0.87 %
	69	68.926552	$\frac{1}{2}^-$	56 m (β^-)			75	74.922522	$\frac{5}{2}^+$	119.8 d (ε)
	70	69.925325	0^+	0.62 %			76	75.919212	0^+	9.0 %
	71	70.927727	$\frac{1}{2}^-$	2.4 m (β^-)			77	76.919913	$\frac{1}{2}^-$	7.6 %
	72	71.926856	0^+	46.5 h (β^-)			78	77.917308	0^+	23.5 %
	73	72.929780	$(\frac{3}{2})^-$	24 s (β^-)			79	78.918498	$\frac{7}{2}^+$	< 0.065 My (β^-)

Z	A	Atomic mass (u)	I^π	Abundance or Half-life		Z	A	Atomic mass (u)	I^π	Abundance or Half-life
	80	79.916520	0^+	49.8%			87	86.908884	$\frac{9}{2}^-$	7.0%
	81	80.917991	$(\frac{1}{2})^-$	18.5 m (β^-)			88	87.905619	0^+	82.6%
	82	81.916698	0^+	9.2%			89	88.907450	$\frac{5}{2}^+$	50.5 d (β^-)
	83	82.919117	$(\frac{9}{2})'$	22.5 m (β^-)			90	89.907738	0^+	28.8 y (β^-)
	84	83.918463	0^+	3.3 m (β^-)			91	90.910187	$(\frac{5}{2})^+$	9.5 h (β^-)
							92	91.910944	0^+	2.7 h (β^-)
Br 35	76	75.924528	1^-	16.1 h (ε)			93	92.913987	$(\frac{7}{2}^+)$	7.4 m (β^-)
	77	76.921378	$\frac{3}{2}^-$	57.0 h (ε)						
	78	77.921144	1^+	6.46 m (ε)		Y 39	84	83.920310	(5^-)	39 m (ε)
	79	78.918336	$\frac{3}{2}^-$	50.69%			85	84.916437	$(\frac{1}{2})^-$	2.7 h (ε)
	80	79.918528	1^+	17.6 m (β^-)			86	85.914893	4^-	14.7 h (ε)
	81	80.916289	$\frac{3}{2}^-$	49.31%			87	86.910882	$\frac{1}{2}^-$	80.3 h (ε)
	82	81.916802	5^-	35.3 h (β^-)			88	87.909508	4^-	106.6 d (ε)
	83	82.915179	$(\frac{3}{2})^-$	2.39 h (β^-)			89	88.905849	$\frac{1}{2}^-$	100%
	84	83.916503	2^-	31.8 m (β^-)			90	89.907152	2^-	64.1 h (β^-)
	85	84.915612	$(\frac{3}{2})^-$	2.9 m (β^-)			91	90.907303	$\frac{1}{2}^-$	58.5 d (β^-)
							92	91.908917	2^-	3.54 h (β^-)
Kr 36	75	74.931029	?	4.3 m (ε)			93	92.909571	$\frac{1}{2}^-$	10.2 h (β^-)
	76	75.925959	0^+	14.8 h (ε)			94	93.911597	2^-	18.7 m (β^-)
	77	76.924610	$\frac{5}{2}^+$	75 m (ε)						
	78	77.920396	0^+	0.356%		Zr 40	87	86.914817	$(\frac{9}{2}^+)$	1.6 h (ε)
	79	78.920084	$\frac{1}{2}^-$	35.0 h (ε)			88	87.910225	0^+	83.4 d (ε)
	80	79.916380	0^+	2.27%			89	88.908890	$\frac{9}{2}^-$	78.4 h (ε)
	81	80.916590	$\frac{7}{2}^+$	0.21 My (ε)			90	89.904703	0^+	51.5%
	82	81.913482	0^+	11.6%			91	90.905644	$\frac{5}{2}^+$	11.2%
	83	82.914135	$\frac{9}{2}^+$	11.5%			92	91.905039	0^+	17.1%
	84	83.911507	0^+	57.0%			93	92.906474	$\frac{5}{2}^+$	1.5 My (β^-)
	85	84.912531	$\frac{9}{2}^+$	10.7 y (β^-)			94	93.906315	0^+	17.4%
	86	85.910616	0^+	17.3%			95	94.908042	$\frac{5}{2}^+$	64.0 d (β^-)
	87	86.913360	$\frac{5}{2}^+$	76 m (β^-)			96	95.908275	0^+	2.80%
	88	87.914453	0^+	2.84 h (β^-)			97	96.910950	$\frac{1}{2}^+$	16.9 h (β^-)
	89	88.917640	$(\frac{3}{2})^+$	3.18 m (β^-)			98	97.912735	0^+	31 s (β^-)
Rb 37	82	81.918195	1^+	1.25 m (ε)		Nb 41	89	88.913449	$(\frac{1}{2})^-$	2.0 h (ε)
	83	82.915144	$\frac{5}{2}^-$	86.2 d (ε)			90	89.911263	8^+	14.6 h (ε)
	84	83.914390	2^-	32.9 d (ε)			91	90.906991	$(\frac{9}{2})^+$	700 y (ε)
	85	84.911794	$\frac{5}{2}^-$	72.17%			92	91.907192	$(7)^+$	35 My (ε)
	86	85.911172	2^-	18.8 d (β^-)			93	92.906377	$\frac{9}{2}^+$	100%
	87	86.909187	$\frac{3}{2}^-$	27.83%			94	93.907281	6^+	0.020 My (β^-)
	88	87.911326	2^-	17.8 m (β^-)			95	94.906835	$\frac{9}{2}^+$	35.0 d (β^-)
	89	88.912278	$(\frac{3}{2})^-$	15.2 m (β^-)			96	95.908100	6^+	23.4 h (β^-)
	90	89.914811	(1^-)	153 s (β^-)			97	96.908097	$\frac{9}{2}^+$	72 m (β^-)
Sr 38	81	80.923270	$(\frac{1}{2}^-)$	22 m (ε)		Mo 42	90	89.913933	0^+	5.67 h (ε)
	82	81.918414	0^+	25.0 d (ε)			91	90.911755	$\frac{9}{2}^+$	15.5 m (ε)
	83	82.917566	$\frac{7}{2}^+$	32.4 d (ε)			92	91.906808	0^+	14.8%
	84	83.913430	0^+	0.56%			93	92.906813	$\frac{5}{2}^+$	3500 y (ε)
	85	84.912937	$\frac{9}{2}^+$	64.8 d (ε)			94	93.905085	0^+	9.3%
	86	85.909267	0^+	9.8%			95	94.905841	$\frac{5}{2}^+$	15.9%

Z	A	Atomic mass (u)	I^π	Abundance or Half-life		Z	A	Atomic mass (u)	I^π	Abundance or Half-life
	96	95.904679	0^+	16.7%			108	107.903895	0^+	26.7%
	97	96.906021	$\frac{5}{2}^+$	9.6%			109	108.905954	$\frac{5}{2}^-$	13.4 h (β^-)
	98	97.905407	0^+	24.1%			110	109.905167	0^+	11.8%
	99	98.907711	$\frac{1}{2}^+$	66.0 h (β^-)			111	110.907660	$\frac{5}{2}^+$	23 m (β^-)
	100	99.907477	0^+	9.6%			112	111.907323	0^+	21.0 h (β^-)
	101	100.910345	$\frac{1}{2}^+$	14.6 m (β^-)						
Tc 43	94	93.909654	7^+	293 m (ε)		Ag 47	103	102.908980	$\frac{7}{2}^+$	65.7 m (ε)
	95	94.907657	$\frac{9}{2}^+$	20.0 h (ε)			104	103.908623	5^+	69.2 m (ε)
	96	95.907870	7^+	4.3 d (ε)			105	104.906520	$\frac{1}{2}^-$	41.3 d (ε)
	97	96.906364	$\frac{9}{2}^+$	2.6 My (ε)			106	105.906662	1^+	24.0 m (ε)
	98	97.907215	$(6)^+$	4.2 My (β^-)			107	106.905092	$\frac{1}{2}^-$	51.83%
	99	98.906254	$\frac{9}{2}^+$	0.214 My (β^-)			108	107.905952	1^+	2.4 m (β^-)
	100	99.907657	1^+	15.8 s (β^-)			109	108.904756	$\frac{1}{2}^-$	48.17%
Ru 44	94	93.911361	0^+	52 m (ε)			110	109.906111	1^+	24.4 s (β^-)
	95	94.910414	$\frac{5}{2}^+$	1.65 h (ε)			111	110.905295	$\frac{1}{2}^-$	7.45 d (β^-)
	96	95.907599	0^+	5.5%			112	111.907010	2^-	3.14 h (β^-)
	97	96.907556	$\frac{5}{2}^+$	2.88 d (ε)		Cd 48	104	103.909851	0^+	58 m (ε)
	98	97.905287	0^+	1.86%			105	104.909459	$\frac{5}{2}^+$	56.0 m (ε)
	99	98.905939	$\frac{5}{2}^+$	12.7%			106	105.906461	0^+	1.25%
	100	99.904219	0^+	12.6%			107	106.906613	$\frac{5}{2}^+$	6.50 h (ε)
	101	100.905582	$\frac{5}{2}^+$	17.0%			108	107.904176	0^+	0.89%
	102	101.904348	0^+	31.6%			109	108.904953	$\frac{5}{2}^+$	463 d (ε)
	103	102.906323	$\frac{3}{2}^+$	39.4 d (β^-)			110	109.903005	0^+	12.5%
	104	103.905424	0^+	18.7%			111	110.904182	$\frac{1}{2}^+$	12.8%
	105	104.907744	$\frac{3}{2}^+$	4.44 h (β^-)			112	111.902757	0^+	24.1%
	106	105.907321	0^+	372 d (β^-)			113	112.904400	$\frac{1}{2}^+$	12.2%
	107	106.910130	$(\frac{5}{2}^+)$	3.8 m (β^-)			114	113.903357	0^+	28.7%
Rh 45	98	97.910716	$(2)^+$	8.7 m (ε)			115	114.905430	$\frac{1}{2}^+$	53.4 h (β^-)
	99	98.908192	$(\frac{1}{2}^-)$	16.1 d (ε)			116	115.904755	0^+	7.5%
	100	99.908116	1^-	20.8 h (ε)			117	116.907228	$\frac{1}{2}^+$	2.4 h (β^-)
	101	100.906159	$\frac{1}{2}^-$	3.3 y (ε)			118	117.911700	0^+	50.3 m (β^-)
	102	101.906814	6^+	2.9 y (ε)		In 49	110	109.907230	2^+	69.1 m (ε)
	103	102.905500	$\frac{1}{2}^-$	100 %			111	110.905109	$\frac{9}{2}^+$	2.83 d (ε)
	104	103.906651	1^+	42.3 s (β^-)			112	111.905536	1^+	14.4 m (ε)
	105	104.905686	$\frac{7}{2}^+$	35.4 h (β^-)			113	112.904061	$\frac{9}{2}^+$	4.3%
	106	105.907279	1^+	29.8 s (β^-)			114	113.904916	1^+	71.9 s (β^-)
Pd 46	99	98.911763	$(\frac{5}{2}^+)$	21.4 m (ε)			115	114.903882	$\frac{9}{2}^+$	95.7%
	100	99.908527	0^+	3.6 d (ε)			116	115.905264	1^+	14.1 s (β^-)
	101	100.908287	$\frac{5}{2}^-$	8.5 h (ε)			117	116.904517	$\frac{9}{2}^+$	43.8 m (β^-)
	102	101.905634	0^+	1.0%		Sn 50	109	108.911294	$\frac{7}{2}^+$	18.0 m (ε)
	103	102.906114	$\frac{5}{2}^+$	17.0 d (ε)			110	109.907858	0^+	4.1 h (ε)
	104	103.904029	0^+	11.0%			111	110.907741	$\frac{7}{2}^+$	35 m (ε)
	105	104.905079	$\frac{5}{2}^+$	22.2%			112	111.904826	0^+	1.01%
	106	105.903478	0^+	27.3%			113	112.905176	$\frac{1}{2}^+$	115.1 d (ε)
	107	106.905127	$\frac{5}{2}^-$	6.5 My (β^-)			114	113.902784	0^+	0.67%
							115	114.903348	$\frac{1}{2}^+$	0.38%

	Z	A	Atomic mass (u)	I^π	Abundance or Half-life
		116	115.901747	0^+	14.6%
		117	116.902956	$\frac{1}{2}^+$	7.75%
		118	117.901609	0^+	24.3%
		119	118.903311	$\frac{1}{2}^+$	8.6%
		120	119.902199	0^+	32.4%
		121	120.904239	$\frac{3}{2}^+$	27.1 h (β^-)
		122	121.903440	0^+	4.56%
		123	122.905722	$\frac{11}{2}^-$	129 d (β^-)
		124	123.905274	0^+	5.64%
		125	124.907785	$\frac{11}{2}^-$	9.62 d (β^-)
		126	125.907654	0^+	0.1 My (β^-)
		127	126.910355	$(\frac{11}{2}^-)$	2.1 h (β^-)
Sb	51	118	117.905534	1^+	3.6 m (ε)
		119	118.903948	$\frac{5}{2}^+$	38.0 h (ε)
		120	119.905077	1^+	15.8 m (ε)
		121	120.903821	$\frac{5}{2}^+$	57.3%
		122	121.905179	2^-	2.70 d (β^-)
		123	122.904216	$\frac{7}{2}^+$	42.7%
		124	123.905938	3^-	60.2 d (β^-)
		125	124.905252	$\frac{7}{2}^+$	2.7 y (β^-)
		126	125.907250	8^-	12.4 d (β^-)
		127	126.906919	$\frac{7}{2}^+$	3.85 d (β^-)
Te	52	117	116.908630	$\frac{1}{2}^+$	62 m (ε)
		118	117.905908	0^+	6.00 d (ε)
		119	118.906411	$\frac{1}{2}^+$	16.0 h (ε)
		120	119.904048	0^+	0.091%
		121	120.904947	$\frac{1}{2}^+$	16.8 d (ε)
		122	121.903050	0^+	2.5%
		123	122.904271	$\frac{1}{2}^+$	0.89%
		124	123.902818	0^+	4.6%
		125	124.904429	$\frac{1}{2}^+$	7.0%
		126	125.903310	0^+	18.7%
		127	126.905221	$\frac{3}{2}^+$	9.4 h (β^-)
		128	127.904463	0^+	31.7%
		129	128.906594	$\frac{3}{2}^+$	69 m (β^-)
		130	129.906229	0^+	34.5%
		131	130.908528	$\frac{3}{2}^+$	25.0 m (β^-)
		132	131.908517	0^+	78.2 h (β^-)
		133	132.910910	$(\frac{3}{2}^+)$	12.5 m (β^-)
I	53	123	122.905594	$\frac{5}{2}^+$	13.2 h (ε)
		124	123.906207	2^-	4.18 d (ε)
		125	124.904620	$\frac{5}{2}^+$	60.2 d (ε)
		126	125.905624	2^-	13.0 d (ε)
		127	126.904473	$\frac{5}{2}^+$	100%
		128	127.905810	1^+	25.0 m (β^-)
		129	128.904986	$\frac{7}{2}^+$	16 My (β^-)
		130	129.906713	5^+	12.4 h (β^-)
		131	130.906114	$\frac{7}{2}^+$	8.04 d (β^-)
		132	131.907987	4^+	2.30 h (β^-)
Xe	54	121	120.911450	$(\frac{5}{2}^+)$	40.1 m (ε)
		122	121.908170	0^+	20.1 h (ε)
		123	122.908469	$(\frac{1}{2}^+)$	2.08 h (ε)
		124	123.905894	0^+	0.096%
		125	124.906397	$(\frac{1}{2}^+)$	17 h (ε)
		126	125.904281	0^+	0.090%
		127	126.905182	$(\frac{1}{2}^+)$	36.4 d (ε)
		128	127.903531	0^+	1.92%
		129	128.904780	$\frac{1}{2}^+$	26.4%
		130	129.903509	0^+	4.1%
		131	130.905072	$\frac{3}{2}^+$	21.2%
		132	131.904144	0^+	26.9%
		133	132.905888	$\frac{3}{2}^+$	5.25 d (β^-)
		134	133.905395	0^+	10.4%
		135	134.907130	$\frac{3}{2}^+$	9.1 h (β^-)
		136	135.907214	0^+	8.9%
		137	136.911557	$\frac{7}{2}^-$	3.82 m (β^-)
Cs	55	130	129.906753	1^+	29.2 m (ε)
		131	130.905444	$\frac{5}{2}^+$	9.69 d (ε)
		132	131.906431	2	6.47 d (ε)
		133	132.905429	$\frac{7}{2}^+$	100%
		134	133.906696	4^+	2.06 y (β^-)
		135	134.905885	$\frac{7}{2}^+$	3 My (β^-)
		136	135.907289	5^+	13.1 d (β^-)
		137	136.907073	$\frac{7}{2}^+$	30.2 y (β^-)
		138	137.911004	3^-	32.2 m (β^-)
Ba	56	127	126.911130	$(\frac{1}{2}^+)$	12.7 m (ε)
		128	127.908237	0^+	2.43 d (ε)
		129	128.908642	$\frac{1}{2}^+$	2.2 h (ε)
		130	129.906282	0^+	0.106%
		131	130.906902	$\frac{1}{2}^+$	12.0 d (ε)
		132	131.905042	0^+	0.101%
		133	132.905988	$\frac{1}{2}^+$	10.7 y (ε)
		134	133.904486	0^+	2.42%
		135	134.905665	$\frac{3}{2}^+$	6.59%
		136	135.904553	0^+	7.85%
		137	136.905812	$\frac{3}{2}^+$	11.2%
		138	137.905232	0^+	71.7%
		139	138.908826	$\frac{7}{2}^-$	82.9 m (β^-)
		140	139.910581	0^+	12.7 d (β^-)
		141	140.914363	$\frac{3}{2}^-$	18.3 m (β^-)

Z	A	Atomic mass (u)	I^π	Abundance or Half-life
La 57	135	134.906953	3/2	19.5 h (ε)
	136	135.907630	1+	9.87 m (ε)
	137	136.906460	7/2	0.06 My (ε)
	138	137.907105	5-	0.089%
	139	138.906347	7/2	99.911%
	140	139.909471	3	40.3 h (β)
	141	140.910896	7/2	3.90 h (β-)
	142	141.914090	2	91.1 m (β)
Ce 58	133	132.911360	1/2+	5.4 h (ε)
	134	133.908890	0+	76 h (ε)
	135	134.909117	1/2+	17.6 h (ε)
	136	135.907140	0+	0.190%
	137	136.907780	3/2+	9.0 h (ε)
	138	137.905985	0+	0.254%
	139	138.906631	3/2+	137.2 d (ε)
	140	139.905433	0+	88.5%
	141	140.908271	7/2-	32.5 d (β)
	142	141.909241	0+	11.1%
	143	142.912383	3/2-	33.0 h (β)
	144	143.913643	0+	284 d (β)
	145	144.917230	3/2-	2.98 m (β)
Pr 59	138	137.910748	1+	1.45 m (ε)
	139	138.908917	5/2+	4.4 h (ε)
	140	139.909071	1+	3.39 m (ε)
	141	140.907647	5/2+	100%
	142	141.910039	2-	19.2 h (β)
	143	142.910814	7/2+	13.6 d (β)
	144	143.913301	0-	17.3 m (β)
Nd 60	139	138.911920	3/2+	29.7 m (ε)
	140	139.909306	0+	3.37 d (ε)
	141	140.909594	3/2+	2.5 h (ε)
	142	141.907719	0+	27.2%
	143	142.909810	7/2	12.2%
	144	143.910083	0+	23.8%
	145	144.912570	7/2-	8.3%
	146	145.913113	0+	17.2%
	147	146.916097	5/2	11.0 d (β)
	148	147.916889	0+	5.7%
	149	148.920145	5/2	1.73 h (β)
	150	149.920887	0+	5.6%
	151	150.923825	(3/2)	12.4 m (β-)
	152	151.924680	0+	11.4 m (β)
Pm 61	142	141.912970	1+	40.5 s (ε)
	143	142.910930	5/2	265 d (ε)
	144	143.912588	5-	349 d (ε)
	145	144.912743	5/2	17.7 y (ε)
	146	145.914708	3-	5.5 y (ε)
	147	146.915135	7/2	2.62 y (β-)
	148	147.917473	1	5.37 d (β-)
	149	148.918332	7/2	53.1 h (β-)
	150	149.920981	(1-)	2.68 h (β-)
Sm 62	142	141.915206	0+	72.5 m (ε)
	143	142.914626	3/2	8.83 m (ε)
	144	143.911998	0+	3.1%
	145	144.913409	7/2	340 d (ε)
	146	145.913053	0+	103 My (α)
	147	146.914894	7/2	15.1%
	148	147.914819	0+	11.3%
	149	148.917180	7/2	13.9%
	150	149.917273	0+	7.4%
	151	150.919929	5/2	90 y (β-)
	152	151.919728	0+	26.6%
	153	152.922094	3/2	46.8 h (β)
	154	153.922205	0+	22.6%
	155	154.924636	3/2	22.4 m (β-)
Eu 63	148	147.918125	5-	54.5 d (ε)
	149	148.917926	5/2	93.1 d (ε)
	150	149.919702	0	36 y (ε)
	151	150.919847	5/2	47.9%
	152	151.921742	3-	13 y (ε)
	153	152.921225	5/2	52.1%
	154	153.922975	3-	8.5 y (β-)
	155	154.922889	5/2	4.9 y (β-)
	156	155.924752	0+	15 d (β)
	157	156.925418	5/2	15 h (β)
Gd 64	149	148.919344	7/2	9.4 d (ε)
	150	149.918662	0+	1.8 My (α)
	151	150.920346	7/2	120 d (ε)
	152	151.919786	0+	0.20%
	153	152.921745	3/2-	242 d (ε)
	154	153.920861	0+	2.1%
	155	154.922618	3/2	14.8%
	156	155.922118	0+	20.6%
	157	156.923956	3/2	15.7%
	158	157.924099	0+	24.8%
	159	158.926384	3/2	18.6 h (β)
	160	159.927049	0+	21.8%
	161	160.929664	5/2	3.7 m (β)
Tb 65	156	155.924742	3	5.34 d (ε)
	157	156.924023	3/2+	150 y (ε)

Z	A	Atomic mass (u)	I^π	Abundance or Half-life
	158	157.925411	3	150 y (ϵ)
	159	158.925342	$\frac{3}{2}$	100 %
	160	159.927163	3	72.1 d (β^-)
	161	160.927566	$\frac{3}{2}$	6.90 d (β^-)
	162	161.929510	1	7.76 m (β^-)
Dy 66	153	152.925769	$\frac{7}{2}$	6.4 h (ϵ)
	154	153.924429	0	3 My (α)
	155	154.925747	$\frac{3}{2}$	10.0 h (ϵ)
	156	155.924277	0	0.057%
	157	156.925460	$\frac{3}{2}$	8.1 h (ϵ)
	158	157.924403	0	0.100 %
	159	158.925735	$\frac{3}{2}$	144.4 d (ϵ)
	160	159.925193	0	2.3%
	161	160.926930	$\frac{5}{2}$	19.90%
	162	161.926795	0	25.5%
	163	162.928728	$\frac{5}{2}$	24.9%
	164	163.929171	0	28.1%
	165	164.931700	$\frac{7}{2}$	2.33 h (β^-)
	166	165.932803	0	81.6 h (β^-)
Ho 67	162	161.929092	1	15 m (ϵ)
	163	162.928731	$(\frac{7}{2})$	33 y (ϵ)
	164	163.930285	1	29.0 m (ϵ)
	165	164.930319	$\frac{7}{2}$	100 %
	166	165.932281	0	26.8 h (β^-)
	167	166.933127	$(\frac{7}{2})$	3.1 h (β^-)
Er 68	160	159.929080	0	28.6 h (ϵ)
	161	160.929996	$\frac{3}{2}$	3.24 h (ϵ)
	162	161.928775	0	0.14%
	163	162.930030	$\frac{5}{2}$	75.1 m (ϵ)
	164	163.929198	0	1.56%
	165	164.930723	$\frac{5}{2}$	10.4 h (ϵ)
	166	165.930290	0	33.4%
	167	166.932046	$\frac{7}{2}$	22.9%
	168	167.932368	0	27.1%
	169	168.934588	$\frac{1}{2}$	9.40 d (β^-)
	170	169.935461	0	14.9%
	171	170.938027	$\frac{5}{2}$	7.52 h (β^-)
	172	171.939353	0	49.3 h (β^-)
Tm 69	166	165.933561	2	7.70 h (ϵ)
	167	166.932848	$\frac{1}{2}$	9.25 d (ϵ)
	168	167.934170	3	93.1 d (ϵ)
	169	168.934212	$\frac{1}{2}$	100 %
	170	169.935798	1	128.6 d (β^-)
	171	170.936427	$\frac{1}{2}$	1.92 y (β^-)
	172	171.938397	2	63.6 h (β^-)

Z	A	Atomic mass (u)	I^π	Abundance or Half-life
Yb 70	166	165.933875	0	56.7 h (ϵ)
	167	166.934946	$\frac{5}{2}$	17.5 m (ϵ)
	168	167.933894	0	0.135%
	169	168.935186	$\frac{7}{2}$	32.0 d (ϵ)
	170	169.934759	0	3.1%
	171	170.936323	$\frac{1}{2}$	14.4%
	172	171.936378	0	21.9%
	173	172.938208	$\frac{5}{2}$	16.2%
	174	173.938859	0	31.6%
	175	174.941273	$\frac{7}{2}$	4.19 d (β^-)
	176	175.942564	0	12.6%
	177	176.945253	$\frac{9}{2}$	1.9 h (β^-)
	178	177.946639	0	74 m (β^-)
Lu 71	172	171.939085	(4)	6.70 d (ϵ)
	173	172.938929	$\frac{7}{2}$	1.37 y (ϵ)
	174	173.940336	1	3.3 y (ϵ)
	175	174.940770	$\frac{7}{2}$	97.39%
	176	175.942679	7	2.61%
	177	176.943752	$\frac{7}{2}$	6.71 d (β^-)
	178	177.945963	1	28.4 m (β^-)
Hf 72	171	170.940490	$(\frac{7}{2})$	12.1 h (ϵ)
	172	171.939460	0	1.87 y (ϵ)
	173	172.940650	$\frac{1}{2}$	24.0 h (ϵ)
	174	173.940044	0	0.16%
	175	174.941507	$\frac{5}{2}$	70 d (ϵ)
	176	175.941406	0	5.2%
	177	176.943217	$\frac{7}{2}$	18.6%
	178	177.943696	0	27.1%
	179	178.945812	$\frac{9}{2}$	13.7%
	180	179.946546	0	35.2%
	181	180.949096	$\frac{1}{2}$	42.4 d (β^-)
	182	181.950550	0	9 My (β^-)
	183	182.953530	$(\frac{3}{2})$	64 m (β^-)
Ta 73	178	177.945750	1	9.31 m (ϵ)
	179	178.945930	$(\frac{7}{2})$	665 d (ϵ)
	180	179.947462	1	0.0123%
	181	180.947992	$\frac{7}{2}$	99.9877%
	182	181.950149	3	115 d (β^-)
	183	182.951369	$\frac{7}{2}$	5.1 d (β^-)
W 74	178	177.945840	0	21.5 d (ϵ)
	179	178.947067	$(\frac{7}{2})$	38 m (ϵ)
	180	179.946701	0	0.13%
	181	180.948192	$\frac{9}{2}$	121 d (ϵ)
	182	181.948202	0	26.3%
	183	182.950220	$\frac{1}{2}$	14.3%

Z	A	Atomic mass (u)	I^π	Abundance or Half-life		Z	A	Atomic mass (u)	I^π	Abundance or Half-life
	184	183.950928	0+	30.7%			198	197.967869	0+	7.2%
	185	184.953416	3/2	75.1 d (β-)			199	198.970552	(5/2)	30.8 m (β-)
	186	185.954357	0+	28.6%			200	199.971417	0+	12.5 h (β-)
	187	186.957153	3/2	23.9 h (β-)		Au 79	194	193.965348	1-	39.5 h (ε)
	188	187.958480	0+	69.4 d (β-)			195	194.965013	3/2+	186 d (ε)
Re 75	182	181.951210	2+	12.7 h (ε)			196	195.966544	2-	6.18 d (ε)
	183	182.950817	(5/2)+	71 d (ε)			197	196.966543	3/2+	100%
	184	183.952530	3-	38 d (ε)			198	197.968217	2-	2.696 d (β-)
	185	184.952951	5/2+	37.40%			199	198.968740	3/2+	3.14 d (β-)
	186	185.954984	1-	90.6 h (β-)			200	199.970670	1-	48.4 m (β-)
	187	186.955744	5/2+	62.60%		Hg 80	193	192.966560	3/2	3.8 h (ε)
	188	187.958106	1-	16.9 h (β-)			194	193.965391	0+	520 y (ε)
	189	188.959219	(5/2)+	24.3 h (β-)			195	194.966640	1/2	9.5 h (ε)
Os 76	182	181.952120	0+	21.5 h (ε)			196	195.965807	0+	0.15%
	183	182.953290	(9/2)+	13.0 h (ε)			197	196.967187	1/2	64.1 h (ε)
	184	183.952488	0+	0.018%			198	197.966743	0+	10.0%
	185	184.954041	1/2	93.6 d (ε)			199	198.968254	1/2	16.8%
	186	185.953830	0+	1.6%			200	199.968300	0+	23.1%
	187	186.955741	1/2	1.6%			201	200.970277	3/2	13.2%
	188	187.955830	0+	13.3%			202	201.970617	0+	29.8%
	189	188.958137	3/2	16.1%			203	202.972848	5/2	46.6 d (β-)
	190	189.958436	0+	26.4%			204	203.973467	0+	6.9%
	191	190.960920	9/2	15.4 d (β-)			205	204.976047	1/2	5.2 m (β-)
	192	191.961467	0+	41.0%		Tl 81	200	199.970934	2-	26.1 h (ε)
	193	192.964138	3/2	30.6 h (β-)			201	200.970794	1/2+	73 h (ε)
	194	193.965173	0+	6.0 y (β-)			202	201.972085	2-	12.2 d (ε)
Ir 77	188	187.958830	(2-)	41.5 h (ε)			203	202.972320	1/2+	29.5%
	189	188.958712	3/2+	13.1 d (ε)			204	203.973839	2-	3.77 y (β-)
	190	189.960580	(4+)	11.8 d (ε)			205	204.974401	1/2+	70.5%
	191	190.960584	3/2+	37.3%			206	205.976084	0-	4.20 m (β-)
	192	191.962580	4-	74.2 d (β-)		Pb 82	201	200.972830	5/2	9.3 h (ε)
	193	192.962917	3/2+	62.7%			202	201.972134	0+	0.05 My (ε)
	194	193.965069	1-	19.2 h (β-)			203	202.973365	5/2	51.9 h (ε)
	195	194.965966	(3/2-)	2.8 h (β-)			204	203.973020	0+	1.42%
Pt 78	187	186.960470	3/2	2.35 h (ε)			205	204.974458	5/2-	15 My (ε)
	188	187.959386	0+	10.2 d (ε)			206	205.974440	0+	24.1%
	189	188.960817	3/2	10.9 h (ε)			207	206.975872	1/2	22.1%
	190	189.959917	0+	0.013%			208	207.976627	0+	52.3%
	191	190.961665	3/2	2.9 d (ε)			209	208.981065	9/2+	3.25 h (β-)
	192	191.961019	0+	0.78%			210	209.984163	0+	22.3 y (β-)
	193	192.962977	(1/2-)	50 y (ε)			211	210.988735	(9/2+)	36.1 m (β-)
	194	193.962655	0+	32.9%			212	211.991871	0+	10.6 h (β-)
	195	194.964766	1/2	33.8%		Bi 83	206	205.978478	6-	6.24 d (ε)
	196	195.964926	0+	25.3%			207	206.978446	9/2-	32 y (ε)
	197	196.967315	1/2-	18.3 h (β-)						

Z	A	Atomic mass (u)	I^π	Abundance or Half-life
	208	207.979717	(5^-)	0.368 My (ε)
	209	208.980374	$9/2^-$	100 %
	210	209.984095	1^-	5.01 d (β⁻)
	211	210.987255	$9/2^-$	2.15 m (α)
	212	211.991255	1^-	60.6 m (β⁻)
Po 84	206	205.980456	0^+	8.8 d (ε)
	207	206.981570	$5/2^-$	5.8 h (ε)
	208	207.981222	0^+	2.90 y (α)
	209	208.982404	$1/2^-$	102 y (α)
	210	209.982848	0^+	138.4 d (α)
	211	210.986627	$9/2^+$	0.52 s (α)
At 85	208	207.986510	6^+	1.63 h (ε)
	209	208.986149	$9/2^-$	5.4 h (ε)
	210	209.987126	5^+	8.3 h (ε)
	211	210.987469	$9/2^-$	7.21 h (ε)
	212	211.990725	(1^-)	0.31 s (α)
	213	212.992911	$9/2^-$	0.11 μs (α)
Rn 86	207	206.990690	$5/2^-$	9.3 m (ε)
	210	209.989669	0^+	2.4 h (α)
	211	210.990576	$1/2^-$	14.6 h (ε)
	212	211.990697	0^+	24 m (α)
	218	218.005580	0^+	35 ms (α)
	222	222.017571	0^+	3.82 d (α)
	224		0^-	107 m (β⁻)
Fr 87	209	208.995870	$9/2^-$	50 s (α)
	212	211.996130	5^+	20 m (ε)
	215	215.000310	$9/2^-$	0.12 μs (α)
	220	220.012293	1	27.4 s (α)
	223	223.019733	$(3/2)$	21.8 m (β⁻)
Ra 88	222	222.015353	0^+	38 s (α)
	223	223.018501	$1/2^+$	11.4 d (α)
	224	224.020186	0^+	3.66 d (α)
	225	225.023604	$(1/2)^+$	14.8 d (β⁻)
	226	226.025403	0^+	1602 y (α)
	227	227.029171	$(3/2^+)$	42 m (β⁻)
Ac 89	224	224.021685	(0^-)	2.9 h (ε)
	225	225.023205	$(3/2^-)$	10.0 d (α)
	226	226.026084	(1^-)	29 h (β⁻)
	227	227.027750	$3/2^-$	21.77 y (β⁻)
	228	228.031015	(3^+)	6.1 h (β⁻)
Th 90	228	228.028715	0^+	1.91 y (α)
	229	229.031755	$5/2^+$	7300 y (α)
	230	230.033128	0^+	75,400 y (α)
	231	231.036299	$5/2^+$	25.52 h (β⁻)

Z	A	Atomic mass (u)	I^π	Abundance or Half-life
	232	232.038051	0^+	100 %
	233	233.041577	$(1/2^+)$	22.3 m (β⁻)
Pa 91	229	229.032073	$(5/2)$	1.4 d (ε)
	230	230.034527	(2^-)	17.7 d (ε)
	231	231.035880	$3/2^-$	32,800 y (α)
	232	232.038565	(2^-)	1.31 d (β⁻)
	233	233.040243	$3/2^-$	27.0 d (β⁻)
U 92	233	233.039628	$5/2^+$	0.1592 My (α)
	234	234.040947	0^+	0.245 My (α)
	235	235.043924	$7/2^-$	0.720%
	236	236.045563	0^+	23.42 My (α)
	237	237.048725	$1/2^+$	6.75 d (β⁻)
	238	238.050785	0^+	99.275%
	239	239.054290	$5/2^+$	23.5 m (β⁻)
Np 93	236	236.046550	(6^-)	0.11 My (ε)
	237	237.048168	$5/2^+$	2.14 My (α)
	238	238.050941	2^+	2.117 d (β⁻)
	239	239.052933	$5/2^+$	2.36 d (β⁻)
Pu 94	237	237.048401	$7/2^-$	45.3 d (ε)
	238	238.049555	0^+	87.74 y (α)
	239	239.052158	$1/2^+$	24,100 y (α)
	240	240.053808	0^+	6570 y (α)
	241	241.056846	$5/2^+$	14.4 y (β⁻)
	242	242.058737	0^+	0.376 My (α)
	243	243.061998	$7/2^+$	4.96 h (β⁻)
Am 95	240	240.055278	(3^-)	50.9 h (ε)
	241	241.056824	$5/2^-$	433 y (α)
	242	242.059542	1^-	16.0 h (β⁻)
	243	243.061375	$5/2^-$	7370 y (α)
	244	244.064279	(6^-)	10.1 h (β⁻)
Cm 96	246	246.067218	0^+	4700 y (α)
	247	247.070347	$9/2^-$	16 My (α)
	248	248.072343	0^+	0.34 My (α)
	249	249.075948	$1/2^+$	64 m (β⁻)
Bk 97	246	246.068720	2^-	1.8 d (ε)
	247	247.070300	$(3/2^-)$	1380 y (α)
Cf 98	251	251.079580	$1/2^+$	898 y (α)
	252	252.081621	0^+	2.64 y (α)
Es 99	252	252.082944	$(4^+,5^-)$	472 d (α)
	253	253.084818	$7/2^+$	20.5 d (α)

Z	A	Atomic mass (u)	I^π	Abundance or Half-life		Z	A	Atomic mass (u)	I^π	Abundance or Half-life
Fm 100	256	256.091767	0^+	2.63 h (f)		Lr 103	260	260.105320		180 s (α)
	257	257.095099	$(\frac{9}{2}^+)$	100 d (α)		Rf 104	261	261.108690		65 s (α)
Md 101	257	257.095580	$(\frac{7}{2}^-)$	5.2 h (ϵ)		Ha 105	261	261.111820		1.8 s (α)
	258	258.098570	(8^-)	55 d (α)			262	262.113760		34 s (f)
						106	263	263.118220		0.8 s (f)
No 102	258	258.098150	0^+	1.2 ms (f)						
	259	259.100931	$(\frac{9}{2}^+)$	60 m (α)		107	262	262.122930		115 ms (α)

INDEX

www.ingramcontent.com/pod-product-compliance
Lightning Source LLC
Chambersburg PA
CBHW030003190526
45157CB00014B/167